# PREFACE

This book is a complete guide for the student preparing for the GCSE examination in Physics. It is primarily intended for use during the eight months or so prior to the examination, but may also prove useful at other times during the course.

*Revise Physics* differs in many respects from a conventional text book. First, it is based on a thorough analysis of the Physics syllabuses of the GCSE examining groups, and unlike a conventional text book, informs the student of the scope of a particular syllabus for which he or she is preparing. Secondly, the subject matter consists of a detailed and concisely expressed account of Physics to this level in a form most suitable for revision. A further feature of this book is the inclusion of a large number of diagrams, for so much Physics which is difficult to explain in words can be shown clearly in diagrams.

Following units on the subject areas to be studied there is a self-test section containing many multiple choice questions. These are designed to enable the student to judge his or her progress; answers are provided. This is followed by a section on examination technique, which contains advice for the student on how to tackle the various types of examination papers and the questions they contain. After this there is a large selection of multiple choice, short answer, structured and longer examination questions, most of which are taken from past GCSE examination papers. Each longer question is followed by an outline of the physical principles involved in answering it.

I would like to thank the staff of Charles Letts & Co Ltd for their valuable advice and unfailing courtesy.

Michael Shepherd 1989

# CONTENTS

Introduction vii
   Guide to using this book vii
   Guide to studying and revising viii
   The GCSE viii
   Table of analysis of examination syllabuses ix
   Examination boards: addresses xiv

## 1 Measurements 1
**1.1** Length 1
**1.2** Mass 1
**1.3** Area and volume 1
**1.4** Time 1

## 2 Speed, velocity and acceleration 2
**2.1** Ticker tape vibrator or 'ticker-timer' 2
**2.2** Motion 2
**2.3** Speed 3
**2.4** Velocity 3
**2.5** Acceleration 3
**2.6** Uniformly accelerated motion 4
**2.7** Distance travelled 5
**2.8** Experiments to show uniformly accelerated motion 5
**2.9** Projectiles 7
**2.10** Summary 7

## 3 Force, momentum, work, energy and power 9
**3.1** Newton's Laws of Motion 9
**3.2** The acceleration produced by a force 10
**3.3** Weight 10
**3.4** Motion in a circle 11
**3.5** Momentum 12
**3.6** Action and reaction 12
**3.7** Work 13
**3.8** Energy 14
**3.9** Potential energy 14
**3.10** Kinetic energy 14
**3.11** Power 15
**3.12** Scalar and vector quantities 15
**3.13** Summary 15

## 4 Turning forces and machines 17
**4.1** Experiment to study moments 17
**4.2** Centre of mass 17
**4.3** Location of the centre of mass 18
**4.4** Stability 18
**4.5** Machines 18
**4.6** Work done by a machine: efficiency 18
**4.7** The lever 19
**4.8** Pulleys 19
**4.9** Summary 20

## 5 Density 21
**5.1** Density measurements 21

## 6 Pressure 23
**6.1** Pressure in a liquid or gas (a fluid) 23
**6.2** Atmospheric pressure 24
**6.3** The manometer 24
**6.4** Simple barometer 25
**6.5** Aneroid barometer 25
**6.6** Summary 26

## 7 Forces between molecules in solids 27
**7.1** Elasticity 27
**7.2** Summary 28

# REVISE
# PHYSICS

## A COMPLETE REVISION COURSE FOR
## GCSE

Michael Shepherd BSc, PhD, MInstP
Senior Science Master, Malvern College, Worcestershire

Charles Letts and Co Ltd
London, Edinburgh & New York

First published 1979
by Charles Letts & Co Ltd
Diary House, Borough Road, London SE1 1DW

Revised 1981, 1983, 1986, 1987, 1989
Reprinted 1988

Illustrations: Tek-Art

© Michael Shepherd 1979, 1981, 1983, 1986, 1987, 1989
© Illustrations: Charles Letts & Co Ltd 1979, 1981, 1983, 1986, 1987, 1989

All our Rights Reserved. No part of this
publication may be reproduced, stored in
a retrieval system, or transmitted, in any
form or by any means, electronic, mechanical,
photocopying, recopying, recording or otherwise, without
the prior permission of Charles Letts Publishers

**British Library Cataloguing in Publication Data**
Shepherd, Michael 1937—
Revise Physics: a complete revision course
for GCSE.—5th ed.   (Letts study aids)
1. Physics
I. Title
530   QC23

ISBN 0 85097 787 8


'Letts' is a registered trademark
of Charles Letts (Scotland) Ltd

Printed and bound in Great Britain by
Charles Letts (Scotland) Ltd

## 8 Kinetic theory of matter — 29
8.1 Atoms and molecules — 29
8.2 The states of matter — 29
8.3 Brownian motion — 30
8.4 Diffusion — 31
8.5 Randomness and mean free path — 31
8.6 Summary — 32

## 9 Expansion of solids and liquids — 33
9.1 Solids — 33
9.2 Liquids — 34
9.3 Thermometers — 34
9.4 Summary — 35

## 10 The Behaviour of gases — 36
10.1 Boyle's law — 36
10.2 Charles' law — 37
10.3 Pressure law — 38
10.4 The universal gas law — 39
10.5 Models of a gas — 40
10.6 Summary — 41

## 11 Specific heat capacity — 42
11.1 To measure the specific heat capacity of a solid — 42
11.2 To measure the specific heat capacity of a liquid — 44
11.3 Summary — 44

## 12 Latent heat — 45
12.1 Vaporization — 45
12.2 Fusion — 45
12.3 To measure the specific latent heat of fusion of ice — 45
12.4 To measure the specific latent heat of vaporization of water — 46
12.5 Evaporation — 46
12.6 Summary — 46

## 13 Transmission of heat — 47
13.1 Conduction — 47
13.2 Comparison of thermal conductivities — 47
13.3 Radiation — 47
13.4 Convection — 49
13.5 Summary — 49

## 14 Introduction to waves — 50
14.1 Progressive waves — 50
14.2 The ripple tank — 50
14.3 Reflection — 52
14.4 Refraction — 52
14.5 Interference — 53
14.6 Diffraction — 54
14.7 Summary — 54

## 15 The passage and reflection of light — 56
15.1 Rectilinear propagation — 56
15.2 Reflection at a plane surface — 58
15.3 Summary — 59

## 16 Refraction of light — 60
16.1 Refraction at a plane boundary — 60
16.2 Internal reflection and critical angle — 62
16.3 Refraction at a spherical boundary — 63
16.4 To determine the focal length of a converging (convex) lens — 64
16.5 Construction of ray diagrams — 64
16.6 Simple microscope (magnifying glass) — 66
16.7 The camera — 66
16.8 The eye — 66
16.9 Summary — 67

## 17 The wave behaviour of light — 68
17.1 Diffraction — 68
17.2 Interference — 68
17.3 The diffraction grating — 69
17.4 The electromagnetic spectrum — 69
17.5 Summary — 70

## 18 Sound — 71
18.1 Echoes — 72
18.2 Pitch — 72
18.3 Intensity and loudness — 72
18.4 Quality — 72
18.5 Summary — 73

## 19 Magnets — 74
19.1 Making magnets — 74
19.2 Magnetic fields — 75
19.3 Summary — 76

## 20 Electrostatics — 77
20.1 The gold-leaf electroscope — 77
20.2 Lines of force — 79
20.3 Summary — 79

## 21 Current electricity — 80
21.1 Potential difference — 80
21.2 Electromotive force — 80
21.3 Resistance — 80
21.4 Ammeters and voltmeters — 81
21.5 Resistors in series — 81

| | | | | |
|---|---|---|---|---|
| 21.6 | Resistors in parallel | 82 | 24.7 The p–n junction diode | 99 |
| 21.7 | Energy | 82 | 24.8 The light emitting diode (LED) | 100 |
| 21.8 | Power | 83 | 24.9 The transistor | 100 |
| 21.9 | Cost | 83 | 24.10 Electronic systems | 101 |
| 21.10 | House electrical installation | 84 | 24.11 Logic gates | 101 |
| 21.11 | Fuses | 84 | 24.12 Summary | 102 |
| 21.12 | Earthing | 85 | | |
| 21.13 | Summary | 85 | | |

## 22 Electromagnetism 86

| 22.1 | The electromagnet | 87 |
|---|---|---|
| 22.2 | The electric bell | 87 |
| 22.3 | The force on charges moving in a magnetic field | 88 |
| 22.4 | The d.c. electric motor | 89 |
| 22.5 | The moving coil meter (galvanometer) | 90 |
| 22.6 | Summary | 90 |

## 23 Electromagnetic induction 91

| 23.1 | Laws | 91 |
|---|---|---|
| 23.2 | The simple d.c. dynamo | 91 |
| 23.3 | The simple a.c. dynamo | 92 |
| 23.4 | Alternating current | 92 |
| 23.5 | The transformer | 93 |
| 23.6 | Power transmission | 93 |
| 23.7 | Summary | 94 |

## 24 Electron beams 95

| 24.1 | Thermionic emission | 95 |
|---|---|---|
| 24.2 | The diode | 95 |
| 24.3 | Cathode rays (electron beams) | 96 |
| 24.4 | The effect of electric and magnetic fields | 97 |
| 24.5 | Cathode ray oscilloscope | 97 |
| 24.6 | Semiconductor materials | 98 |

## 25 Radioactivity and atomic structure 103

| 25.1 | Radiation detectors | 103 |
|---|---|---|
| 25.2 | Atomic structure | 104 |
| 25.3 | Isotopes | 104 |
| 25.4 | Radioactivity | 105 |
| 25.5 | Radioactive decay | 105 |
| 25.6 | Safety | 106 |
| 25.7 | Nuclear energy | 106 |
| 25.8 | Summary | 107 |

## Self-test questions 109

## Hints for candidates taking Physics examinations 119

## Practice in answering examination questions 121

| Multiple choice questions | 121 |
|---|---|
| Questions requiring short or structured answers | 137 |
| Longer questions | 194 |

## Internal assessment of practical work 216

## Index 217

# INTRODUCTION

Knowing how to prepare for an examination can be difficult. *Revise Physics* will make it easier by providing the GCSE Physics student with a complete revision course. This book not only contains all the necessary topics, but has a self-test section and an extensive selection of examination questions with answers. Each longer question is keyed back to the relevant unit in the book and is followed by hints on how to answer it. To get the best from the study scheme in this book you are advised to follow the procedure outlined below.

## Guide to using this book

### The table of analysis of examination syllabuses

Turn to page ix for the table of analysis. The Examining Groups (and their alternative syllabuses where relevant) are listed across the top of the page. The contents of this book are listed by unit and sub-unit headings down the left-hand side of the page.

Find the column containing the information about the syllabus you are studying. (Remember, many Examining Groups have alternative Physics syllabuses, so if you are in doubt about the requirements of your particular Group, please check with your teacher – the table is intended only as a useful guide.) In the column you will find the following information:

1 The number of examination papers set with the time allowed in brackets.
2 The percentage of the total mark allocated for teacher assessment of practical work. (This is the mark awarded by your teacher for certain assessed practical tests set during the course, see p. 216).
3 Whether multiple choice, short, structured and longer questions are set. The percentage of the total mark allocated to these different types of question is given.
4 The Physics content of the GCSE courses – the most important information. This information has been divided into twenty-five units, within which are a varying number of sub-units.

Not all these sub-units need to be studied for any one syllabus. The key to the symbols in the columns is:

- • Unit required for a syllabus.
- • (h) Unit required only for the additional or alternative paper.
- ○ Required for option only.

A, B or C Unit required for option A, option B or option C of the MEG (Nuffield) syllabus.

A blank space indicates that a unit is not required.

### The units covering the subject areas to be studied

You are advised to work through as many of the required units as possible. You may find it useful to consult your Examining Group's syllabus as it may give valuable information in the form of additional notes. You can write to the Examining Groups to buy syllabuses. A list of their addresses is given on page xiv.

### Testing yourself and revising

When you feel you have mastered as many units as you can, you should test what you have learned. A self-test section, consisting of multiple choice questions, has been included to enable you to judge your own progress. If you fail to get the majority of these right, go back to the appropriate unit and revise more thoroughly.

### Help with examination technique

This part of the book gives you advice on tackling the different kinds of examination questions and how to do your best under examination conditions.

### Practice in answering examination questions

A large selection of questions has been included to give you maximum practice in answering all types of questions. For convenience these questions have been grouped by type. Be careful not to answer questions on sub-units of the subject material which are not in your syllabus.

## Guide to studying and revising

The best way to ensure a good result in your GCSE examination is to make sure you *understand* the various topics in your course *as you encounter them*. Once a topic is understood, revising and remembering it is made so much easier.

Of course, revision during the last few months leading up to your examination is also important. Time is precious, so make sure you use it well. A good method is to devise a revision timetable – make a list of all the topics you have to study and divide your time up accordingly. Try not to be over-optimistic. You can only realistically concentrate well for 30 minutes at a time. Give yourself frequent short breaks to keep up your powers of concentration.

Where you study can make a big difference. Try to work in a quiet, well-lit, well-ventilated room. Ask your family and friends not to disturb you; it is all too easy to become distracted and fall behind with your revision schedule. If you are intending to work in the evenings, start work as early as possible. You will find your study time far more effective when you are fresh.

Revising is, to a large extent, a matter of technique. It is not good enough to simply read through your notes or text book time and time again. There are various methods which can be used to help you revise; no one method in itself is better or worse than the other. It really depends on which you find easiest. After covering a particular subject it is a good idea to highlight the most important points by making brief notes. Such notes are also useful for jogging your memory in the final stages of revision, just before your exam.

Whichever method you choose, develop it and persevere. A good revision technique can make all the difference to your final examination performance.

## The GCSE

GCSE stands for the General Certificate of Secondary Education. It is mainly designed for pupils in the 5th year of secondary education. However, anyone may take it whatever their age. Candidates may study for it at school, college or privately.

The GCSE is administered by groups of examining boards. There are four groups in England (London East Anglia, Midland, Northern and Southern), one in Wales and one in Northern Ireland. A list of all the groups, boards and their addresses is on page xiv.

Monitoring of the GCSE is carried out by the Secondary Examinations and Assessment Council (SEAC). The Council must ensure that:
1 All the examining groups follow the National Criteria.
2 Standards are comparable between different subjects, different examining groups and from year to year.

The National Criteria are nationally agreed guidelines which all GCSE courses and examinations must follow. They are in two parts:
1 General criteria which apply to all subjects.
2 Subject criteria provide the framework for GCSE in each subject.

The National Criteria were drawn up after consultation with parents, schools, teachers' associations, higher education, employers' organizations, local authorities, the Secondary Examinations Council (predecessor of the SEAC) and the Government. They will be reviewed regularly by the Secondary Examinations and Assessments Council and changed if this is found to be necessary.

In Physics the examining groups offer assessment at two levels. These are referred to by different names by the various examining groups. Generally speaking the lower of the two levels is intended for candidates who are not attempting to gain grades A or B, and the higher one for those who wish to be considered for grades A and B. A candidate should be entered at the level which most suits his or her ability in Physics.

The arrangement of written examination papers varies from one examining group to another. Candidates who wish to be considered for grades A and B either take an additional paper or an alternative to one of the papers. Details for each examining group appear at the head of each column in the syllabus analysis beginning on page xiii. The syllabus analysis table also indicates any topics which are only examined in the additional or alternative paper.

For all Physics syllabuses in the GCSE examination a minimum of 20% of the marks is awarded for teacher assessment of practical work carried out during the course. This takes the form of periodic assessment by the teacher of a number of simple skills in practical work. This assessment will often, but not always, take place in the last year of the course. Further details are given in a separate section on page 216.

The GCSE examination places more emphasis on understanding topics in the syllabus than it does on memorizing information. The examination papers are designed to measure positive achievement, that is to find out what a candidate does know and understand rather than what he or she does not. Fifteen per cent of the marks in the examination are for industrial, economic and social applications of Physics.

# Table of analysis of examination syllabuses

| | | LEAG | | MEG | | NEA | | SEG | | WJEC | NI |
|---|---|---|---|---|---|---|---|---|---|---|---|
| Syllabus | | A(m) | B(a)(m) | (m) | Nuffield (m) | A(m) | B(a)(m) | (m) | alt(a)(m) | (m) | |
| Number of papers (+total time) | | 2(3¼) | 2(3¼) | 2(2) | 2(2) | 1(2) | 1(2) | 2(2) | 2(2) | 1(2½) | 2(2¾) |
| Teacher assessment | | ●20% | ●20% | ●20% | ●28.5% | ●20% | ●20% | ●30% | ●30% | ●20% | ●20% |
| Multiple choice | | ●30% | ●30% | ●40% | ●43% | ●}80% | | ●28% | ●28% | | ●35% |
| Short and structured | | ●50% }50%(b) | ●50% }50%(b) | ●40% | ●28.5% | ● | ●80% | ●42% | ●42% | ●48% | ●30% |
| Longer answers | | ○ | ○ | (c) | (d) | (e) | (f) | (g) | (g) | ●32%(j) | ●15%(k) |
| 1 | Measurements | ● | ● | ● | ● | ● | ● | ● | ● | ● | ● |
| 2.1 | Tickertape vibrator | ● | ● | ● | ● | ● | ● | ● | ● | ● | ● |
| 2.2 | Motion | ● | ● | ● | ● | ● | ● | ● | ● | ● | ● |
| 2.3 | Speed | ● | ● | ● | ● | ● | ● | ● | ● | ● | ● |
| 2.4 | Velocity | ● | ● | ● | ● | ● | ● | ● | ● | ● | ●(2) |
| 2.5 | Acceleration | ● | ● | ● | ● | ● | ● | ● | ● | ● | ●(h) |
| 2.6 | Uniformly accelerated motion | ●(h) | ●(h) | ●(h) | ○(A) | ● | ●(h) | ●(h) | ●(h) | ● | ●(h) |
| 2.7 | Distance travelled | ●(h) | ●(h) | ●(h) | ○(A) | ● | ●(h) | ●(h) | ●(h) | ● | ●(h) |
| 2.8 | Experiments to show uniform acceleration | ●(h) | ● | ●(h) | ● | ● | ● | | ● | | ● |
| 2.9 | Projectiles | | ●(h) | | ●/○(A) | | | | ●(h) | | |
| 3.1 | Newton's Laws of Motion | ● | ● | ● | ● | ● | ● | ● | ●(h) | ● | ●(h) |
| 3.2 | The acceleration produced by a force | ● | ● | ●(h) | ●(h) | ● | ● | ● | ● | ● | ●(h) |
| 3.3 | Weight | ● | ● | ● | ● | ● | ● | ● | ● | ● | ● |
| 3.4 | Motion in a circle | ● | ● | | ○(A) | | | | ●(h) | | |
| 3.5 | Momentum | ● | ● | | ●(2) | | ●(h) | ●(h) | ●(h) | | |
| 3.6 | Action and reaction | ● | ● | ● | ● | ● | ●(h) | | ● | ● | ●(h) |
| 3.7 | Work | ●(h) | ● | ● | ● | ● | ● | ● | | ● | ● |
| 3.8 | Energy | ● | ● | ● | ● | ● | ● | ● | ● | ● | ● |
| 3.9 | Potential energy | ● | ●(2) | ●(2) | ●(2) | ●(h) | ●(h) | ● | | | ●(2) |
| 3.10 | Kinetic energy | ● | ●(2) | ●(2) | ●(2) | ●(h) | ●(h) | ● | | | ●(2) |
| 3.11 | Power | ● | ● | ● | ● | ● | ● | ● | ● | ● | ● |
| 3.12 | Scalar and vector quantities | ● | ●(h) | ●(h) | ● | ● | | ● | ● | ●(h) | ● |
| 4.1 | Experiment to study moments | ● | ● | ● | ● | | | ● | ● | | ● |
| 4.2 | Centre of mass | ● | | | | | | ● | | | ● |
| 4.3 | Location of the centre of mass | ● | | | | | | ● | | | ● |
| 4.4 | Stability | ● | | | | | | ● | | | ● |
| 4.5 | Machines | ● | ● | | | | | ● | ● | | ● |
| 4.6 | Work done by a machine efficiency | ● | ● | ● | ● | ● | | ● | ● | ●(h) | ● |
| 4.7 | The lever | | ● | | | | | ● | ● | | ● |
| 4.8 | Pulleys | | ● | | | | | | | | |
| 5 | Density | ● | ● | ● | ● | ● | ● | ● | ● | ● | ● |
| 5.1 | Density measurements | | ● | ● | ● | | ● | | ● | | ● |
| 6 | Pressure | ● | ● | ● | ● | ● | ● | ● | ● | | ● |

# Table of analysis

| Table of analysis of examination syllabuses *continued* | LEAG A(m) | LEAG B(a)(m) | MEG (m) | MEG Nuffield (m) | NEA A(m) | NEA B(a)(m) | SEG (m) | SEG alt(a)(m) | WJEC (m) | NI |
|---|---|---|---|---|---|---|---|---|---|---|
| Syllabus | A(m) | B(a)(m) | (m) | Nuffield (m) | A(m) | B(a)(m) | (m) | alt(a)(m) | (m) | |
| Number of papers (+ total time) | 2(3¼) | 2(3¼) | 2(2) | 2(2) | 1(2) | 1(2) | 2(2) | 2(2) | 1(2½) | 2(2¾) |
| Teacher assessment | ●20% | ●20% | ●20% | ●28.5% | ●20% | ●20% | ●30% | ●30% | ●20% | ●20% |
| Multiple choice | ●30% | ●30% | ●40% | ●43% | ●}80% | | ●28% | ●28% | | ●35% |
| Short and structured | ●50% }50%(b) | ●50% }50%(b) | ●40% | ●28.5% | ● | ●80% | ●42% | ●42% | ●48% | ●30% |
| Longer answers | ○ | ○ | (c) | (d) | (e) | (f) | (g) | (g) | ●32%(j) | ●15%(k) |
| 6.1 Pressure in a fluid | ● | ● | ● | | ● | ● | ● | ● | | ● |
| 6.2 Atmospheric pressure | ● | ● | ● | ● | ● | ● | ● | ● | | |
| 6.3 The manometer | | | | ● | ● | | ● | ● | | ● |
| 6.4 Simple barometer | ● | ● | | ● | | ● | ● | ● | | |
| 6.5 Aneroid barometer | ● | | | ● | ● | ● | | ● | | |
| 7.1 Elasticity | ● | ● | ● | ● | ● | ● | ● | ● | ● | ● |
| 8.1 Atoms and molecules | ● | ● | ● | ● | ● | ● | ● | ● | ● | ● |
| 8.2 The states of matter | ● | ● | ● | ● | ● | ● | ● | ● | ● | ● |
| 8.3 Brownian motion | ●(h) | ● | ● | ● | ● | ● | ● | ● | ● | ● |
| 8.4 Diffusion | ●(h) | ● | ● | ● | ● | ● | ● | ● | ● | ● |
| 8.5 Randomness and mean free path | ●(h) | | | ● | | | ● | | | ● |
| 9.1 Solids | ● | ● | | ● | ● | ● | ● | ● | ● | ● |
| 9.2 Liquids | ● | ● | ● | ● | ● | ● | ● | ● | ● | ● |
| 9.3 Thermometers | ● | ● | ● | ● | ● | | ● | | ● | ● |
| 10.1 Boyle's law | ● | ● | | ● | ●(2) | | ● | ● | | |
| 10.2 Charles' law | ● | ● | | ● | ●(2) | | ● | | | |
| 10.3 Pressure law | ● | ● | | ● | ●(2) | | ● | | | |
| 10.4 The universal gas law | | | | | | | | | | |
| 10.5 Models of a gas | ● | ● | | ● | ● | ● | | | | |
| 11 Specific heat capacity | | ●(2) | | ●(2) | ●(h) | ● | ● | ● | ● | ●(h) |
| 11.1 The specific heat capacity of a solid | | ●(2) | | ●(2) | | ●(2) | ● | ● | | ●(h) |
| 11.2 The specific heat capacity of a liquid | | ● | | ●(2) | | ●(2) | ● | ● | | ●(h) |
| 12.1 Vaporization | | | ● | ● | ● | ● | ● | ● | ● | ●(h) |
| 12.2 Fusion | | | ●(1) | ●(2) | ● | ● | ● | ● | ● | ●(h) |
| 12.3 The specific latent heat of fusion of ice | | | ● | ●(2) | | ●(1) | | | | ●(h) |
| 12.4 The specific latent heat of vaporization of water | | | ●(1) | ●(2) | | ●(1) | | | | ●(h) |
| 12.5 Evaporation | | ● | ● | ● | ● | ● | ● | ● | ● | |
| 13.1 Conduction | ● | ● | ● | ● | ● | ● | ● | ● | ● | ● |
| 13.2 Comparison of thermal conductivities | ● | ● | ● | ● | ● | | ● | ● | ● | ● |
| 13.3 Radiation | ● | ● | ● | ● | ● | ● | | ● | ● | ● |
| 13.4 Convection | ● | ● | ● | ● | ● | ● | ● | ● | ● | ● |
| 14.1 Progressive waves | ● | ● | ● | ● | ● | ● | ● | ● | ● | ● |
| 14.2 The ripple tank | ● | ● | ● | ● | ● | ● | ● | ● | ● | ● |

# Table of analysis

| Table of analysis of examination syllabuses *continued* | LEAG | | | | MEG | | NEA | | SEG | | WJEC | NI |
|---|---|---|---|---|---|---|---|---|---|---|---|---|
| Syllabus | A(m) | | B(a)(m) | | (m) | Nuffield (m) | A(m) | B(a)(m) | (m) | alt(a)(m) | (m) | |
| Number of papers (+ total time) | 2(3¼) | | 2(3¼) | | 2(2) | 2(2) | 1(2) | 1(2) | 2(2) | 2(2) | 1(2½) | 2(2¾) |
| Teacher assessment | ●20% | | ●20% | | ●20% | ●28.5% | ●20% | ●20% | ●30% | ●30% | ●20% | ●20% |
| Multiple choice | ●30% | | ●30% | | ●40% | ●43% | ●80% | | ●28% | ●28% | | ●35% |
| Short and structured | ●50% | 50% | ●50% | 50% | ●40% | ●28.5% | ● | ●80% | ●42% | ●42% | ●48% | ●30% |
| Longer answers | ○ | (b) | ○ | (b) | (c) | (d) | (e) | (f) | (g) | (g) | ●32%(j) | ●15%(k) |
| 14.3 Reflection | ● | | ● | | ● | ● | ●(h) | ● | ● | ● | ● | |
| 14.4 Refraction | ● | | ● | | ● | ● | ●(h) | ● | | ● | ● | |
| 14.5 Interference | ●(h) | | ● | | ●(1) | ○(B) | | ● | | | ●(h) | |
| 14.6 Diffraction | ● | | ● | | ●(1) | ○(B) | | ● | | ● | ●(h) | |
| 15.1 Rectilinear propagation | ● | | ● | | ● | ● | ● | ● | ● | ● | | ● |
| 15.2 Reflection at a plane surface | ● | | ● | | ● | ● | ● | ● | ● | ● | | ● |
| 16.1 Refraction at a plane boundary | ● | | ● | | ● | ● | ● | ● | ● | | | ● |
| 16.2 Internal reflection and critical angle | ● | | ● | | ● | | ● | ● | ● | | | |
| 16.3 Refraction at a spherical boundary | | | ● | | ● | ● | ● | ● | | ●(h) | | ● |
| 16.4 The focal length of a converging lens | | | ● | | ● | ● | ● | ● | | ● | | ● |
| 16.5 Construction of ray diagrams | ● | | ● | | ● | ● | ● | ● | | ● | | ● |
| 16.6 Simple microscope | ●(h) | | ● | | | ●(h) | ● | | | ● | | |
| 16.7 The camera | ● | | ● | | ● | ● | ● | ● | ● | ● | | |
| 16.8 The eye | | | ● | | | ● | ● | ● | | | | |
| 17.1 Diffraction | | | ● | | ●(1) | ○(B) | | | | | | |
| 17.2 Interference | | | ● | | ●(1) | ○(B) | | | | | | |
| 17.3 The diffraction grating | | | | | | | | | | | | |
| 17.4 The electromagnetic spectrum | ● | | ● | | ● | ●/○(B) | ● | ● | ● | ● | ● | ● |
| 18 Sound | ● | | | | ● | ● | ● | | ● | ● | ● | ● |
| 18.1 Echoes | | | | | ● | ● | ● | | ● | | ● | ● |
| 18.2 Pitch | ●(h) | | | | ● | | ● | | | | | |
| 18.3 Intensity and loudness | ● | | | | ● | | ● | | | | | |
| 18.4 Quality | | | | | ● | | | | | | | |
| 19 Magnets | ● | | | | | ● | ● | ● | | ● | ● | ● |
| 19.1 Making magnets | ● | | | | | | ● | | | | | |
| 19.2 Magnetic fields | ● | | | | | ● | ● | ● | | ● | ● | |
| 20 Electrostatics | ● | | ● | | | ● | ● | ● | ● | ● | ● | ● |
| 20.1 The leaf electroscope | ● | | | | | | ● | | ● | | | |
| 20.2 Lines of force | | | | | | | | | ● | | | |
| 21 Current electricity | ● | | ● | | ●(2) | ● | ● | ● | ● | ● | ● | ● |
| 21.1 Potential difference | ● | | ● | | ● | ● | ● | ● | ● | ● | ● | ● |
| 21.2 Electromotive force | ● | | | | ● | | ● | | ● | | ●(h) | |
| 21.3 Resistance | ● | | ● | | ● | ● | ● | ● | ● | ● | ● | ● |

## Table of analysis

| Table of analysis of examination syllabuses *continued* | LEAG | | | MEG | | NEA | | SEG | | WJEC | NI |
|---|---|---|---|---|---|---|---|---|---|---|---|
| Syllabus | A(m) | B(a)(m) | | (m) | Nuffield (m) | A(m) | B(a)(m) | (m) | alt(a)(m) | (m) | |
| Number of papers (+ total time) | 2(3¼) | 2(3¼) | | 2(2) | 2(2) | 1(2) | 1(2) | 2(2) | 2(2) | 1(2½) | 2(2¾) |
| Teacher assessment | ●20% | ●20% | | ●20% | ●28.5% | ●20% | ●20% | ●30% | ●30% | ●20% | ●20% |
| Multiple choice | ●30% | ●30% | | ●40% | ●43% | ●⎫80% | | ●28% | ●28% | | ●35% |
| Short and structured | ●50%⎫ | ●50%⎫ | | ●40% | ●28.5% | ●⎭ | ●80% | ●42% | ●42% | ●48% | ●30% |
| Longer answers | ○ ⎬50%(b) | ○ ⎬50%(b) | | (c) | (d) | (e) | (f) | (g) | (g) | ●32%(j) | ●15%(k) |
| 21.4 Ammeters and voltmeters | ● | ● | | ● | ● | ● | ● | ● | ● | ● | ● |
| 21.5 Resistors in series | ● | ● | | ● | | ● | ● | ● | ● | ● | ● |
| 21.6 Resistors in parallel | ●(h) | ● | | ● | | ●(h) | ●(1) | ● | ● | | ● |
| 21.7 Energy | ● | ● | | ●(2) | ● | ● | ● | ● | ● | ● | ● |
| 21.8 Power | ● | ● | | ● | ● | ● | ● | ● | ● | ● | ● |
| 21.9 Cost | ● | ● | | ● | ● | ● | ● | ● | ● | ● | ● |
| 21.10 House electrical installation | ● | ● | | ● | ● | ● | | ● | ● | ● | ● |
| 21.11 Fuses | ● | ● | | ● | ● | ● | ● | ● | ● | ● | ● |
| 21.12 Earthing | ● | ● | | ● | ● | ● | ● | ● | ● | ● | ● |
| 22 Electromagnetism | ● | ● | | ● | ● | ● | ● | ● | ● | ● | ● |
| 22.1 The electromagnet | | ● | | | ● | | ● | ● | ● | ● | |
| 22.2 The electric bell | | | | | ● | | ● | | | ● | ● |
| 22.3 The force on charges moving in a magnetic field | ● | ● | | ● | ● | ● | ● | ● | ● | ●(h) | ● |
| 22.4 The d.c. electric motor | ● | ● | | ● | ● | ● | ● | ● | | ● | ● |
| 22.5 The moving coil meter | | | | | ● | ● | | | | | |
| 23.1 Laws of electromagnetic induction | ● | ● | | ● | ● | ● | ● | ● | | ● | ● |
| 23.2 The d.c. dynamo | ● | | | ●(1) | ● | ● | ● | | | ● | |
| 23.3 The a.c. dynamo | | ● | | ●(1) | | ● | ● | ● | ● | ● | |
| 23.4 Alternating current | | | | | ●(2) | | ●(1) | ● | | | ● |
| 23.5 The transformer | ● | ● | | ● | ● | ● | ● | ● | ● | ● | |
| 23.6 Power transmission | | ● | | ●(2) | ●(2) | ●(h) | ● | ● | ● | ● | |
| 24.1 Thermionic emission | | | | ● | ● | | ● | | | | |
| 24.2 The diode | ● | ● | | ● | ● | | ● | ● | | | |
| 24.3 Cathode rays | | ● | | | ● | | ● | | | | ● |
| 24.4 The effect of electric and magnetic fields | | ● | | ● | ● | | ● | | | ●(h) | ● |
| 24.5 The cathode ray oscilloscope | ● | ● | | ● | ● | ● | ● | ● | ● | ● | ● |
| 24.6 Semiconductor materials | ● | ● | | ● | | | | ● | ● | ● | ● |
| 24.7 The p–n junction diode | ● | ● | | ● | | ● | ● | | ● | ● | |
| 24.8 The light emitting diode | | ● | | ● | ● | ● | ● | | | ● | ● |
| 24.9 The transistor | ● | ● | | | | ● | | | | ● | |
| 24.10 Electronic systems | | ● | | ● | ○(C) | ● | ● | | | ● | |
| 24.11 Logic gates | | ● | | ● | ○(C) | ● | | | | ● | ● |
| 25.1 Radiation detectors | | | | ● | | | ● | | ● | | |

| Table of analysis of examination syllabuses *continued* | LEAG | | MEG | | NEA | | SEG | | WJEC | NI |
|---|---|---|---|---|---|---|---|---|---|---|
| Syllabus | A(m) | B(a)(m) | (m) | Nuffield (m) | A(m) | B(a)(m) | (m) | alt(a)(m) | (m) | |
| Number of papers (+ total time) | 2($3\frac{1}{4}$) | 2($3\frac{1}{4}$) | 2(2) | 2(2) | 1(2) | 1(2) | 2(2) | 2(2) | 1($2\frac{1}{2}$) | 2($2\frac{3}{4}$) |
| Teacher assessment | ●20% | ●20% | ●20% | ●28.5% | ●20% | ●20% | ●30% | ●30% | ●20% | ●20% |
| Multiple choice | ●30% | ●30% | ●40% | ●43% | ●⎫ 80% | | ●28% | ●28% | | ●35% |
| Short and structured | ●50% ⎫ 50% (b) | ●50% ⎫ 50% (b) | ●40% | ●28.5% | ●⎭ | ●80% | ●42% | ●42% | ●48% | ●30% |
| Longer answers | ○ | ○ | (c) | (d) | (e) | (f) | (g) | (g) | ●32%(j) | ●15%(k) |
| 25.2 Atomic structure | ● | ● | ● | ● | ● | ● | ● | ● | ● | ● |
| 25.3 Isotopes | ●(h) | ●(h) | | ● | ● | ●(h) | ● | ● | | |
| 25.4 Radioactivity | ● | ● | ● | ● | ● | ● | ● | ● | ● | ● |
| 25.5 Radioactive decay | ● | ● | ● | ● | ● | ● | ● | ● | ● | ● |
| 25.6 Safety | ● | ● | ● | ● | ● | ● | ● | ● | | ● |
| 25.7 Nuclear energy | | ● | ● | ● | ● | ● | | ● | | |

(a) A Nuffield-based course.

(b) An alternative paper to Paper 2 (short and structured questions) may be taken. This consists of structured and longer questions and is designed for those candidates likely to gain grades A to D.

(c) An additional paper ($1\frac{1}{2}$ hours) consisting of structured and longer questions may be taken. This is intended to differentiate those candidates likely to gain grades A and B.

(d) An additional paper (2 hours) consisting of structured and longer questions may be taken. This is intended to differentiate those candidates likely to gain grades A and B.

(e) An alternative paper ($2\frac{1}{2}$ hours) may be taken. This may include multiple choice and longer questions as well as short answer questions. It is designed for those candidates who aspire to the higher grades.

(f) An alternative paper ($2\frac{1}{2}$ hours) may be taken. This may include longer questions as well as short answer questions. It is designed for those candidates who aspire to higher grades.

(g) An additional paper ($1\frac{1}{4}$ hours) consisting of structured and longer questions may be taken. This is intended to differentiate those candidates likely to gain grades A and B. For those who take this paper the allocation of marks will be approximately 30%, 20%, 25% and 25%.

(h) Only required for the additional or alternative paper.

(j) An alternative paper ($2\frac{1}{2}$ hours) may be taken. This consists of short and structured questions and some longer questions and is designed for those candidates who aspire to grades A and B.

(k) An additional paper (2 hours) consisting of short and longer questions may be taken. This is intended to differentiate between those candidates likely to gain grades A and B. For those who take this paper the allocation of marks will be 10%, 17.5%, 40% and 32.5%.

(m) Some useful equations are printed on the examination paper or on an accompanying sheet.

(A), (B) and (C) refer to options in the MEG (Nuffield) syllabus.

(1) Qualitative only

(2) Qualitative only *except* in the additional or higher paper.

# Examination Boards: Addresses

## NORTHERN EXAMINING ASSOCIATION (NEA)

*JMB*    Joint Matriculation Board
Devas Street, Manchester M15 6EU

*ALSEB*    Associated Lancashire Schools Examining Board
12 Harter Street, Manchester M1 6HL

*NREB*    Northern Regional Examinations Board
Wheatfield Road, Westerhope, Newcastle upon Tyne NE5 5JZ

*NWREB*    North-West Regional Examinations Board
Orbit House, Albert Street, Eccles, Manchester M30 0WL

*YHREB*    Yorkshire and Humberside Regional Examinations Board
Harrogate Office—31-33 Springfield Avenue, Harrogate HG1 2HW
Sheffield Office—Scarsdale House, 136 Derbyshire Lane, Sheffield S8 8SE

## MIDLAND EXAMINING GROUP (MEG)

*Cambridge*    University of Cambridge Local Examinations Syndicate
Syndicate Buildings, 1 Hills Road, Cambridge CB1 2EU

*O & C*    Oxford and Cambridge Schools Examination Board
10 Trumpington Street, Cambridge CB2 1QB *and* Elsfield Way, Oxford OX2 8EP

*SUJB*    Southern Universities' Joint Board for School Examinations
Cotham Road, Bristol BS6 6DD

*WMEB*    West Midlands Examinations Board
Norfolk House, Smallbrook Queensway, Birmingham B5 4NJ

*EMREB*    East Midland Regional Examinations Board
Robins Wood House, Robins Wood Road, Aspley, Nottingham NG8 3NR

## LONDON EAST ANGLIAN GROUP (LEAG)

*London*    University of London Schools Examinations Board
Stewart House, 32 Russell Square, London WC1B 5DN

*LREB*    London Regional Examining Board
Lyon House, 104 Wandsworth High Street, London SW18 4LF

*EAEB*    East Anglian Examinations Board
'The Lindens', Lexden Road, Colchester CO3 3RL

## SOUTHERN EXAMINING GROUP (SEG)

*AEB*    The Associated Examining Board
Stag Hill House, Guildford GU2 5XJ

*Oxford*    Oxford Delegacy of Local Examinations
Ewert Place, Summertown, Oxford OX2 7BZ

*SREB*    Southern Regional Examinations Board
Eastleigh House, Market Street, Eastleigh, Southampton SO5 4SW

*SEREB*    South-East Regional Examinations Board
Beloe House, 2-10 Mount Ephraim Road, Tunbridge Wells TN1 1EU

*SWEB*    South-Western Examinations Board
23-29 Marsh Street, Bristol BS1 4BP

## WALES

*WJEC*    Welsh Joint Education Committee
245 Western Avenue, Cardiff CF5 2YX

## NORTHERN IRELAND

*NISEC*    Northern Ireland Schools Examinations Council
Beechill House, 42 Beechill Road, Belfast BT8 4RS

## SCOTLAND

*SEB*    Scottish Examination Board
Ironmills Road, Dalkeith, Midlothian EH22 1BR

# 1 MEASUREMENTS

All measurements in Physics are related to the three chosen fundamental quantities of **length**, **mass** and **time**. For many years scientists have agreed to use the metric system; the particular one used now is based on the metre, the kilogram and the second.

## 1.1 Length

The unit of length is the metre. Various multiples or submultiples are also used. Thus

$$1 \text{ kilometre (km)} = 1000 \text{ metres (m)}$$
$$1 \text{ metre} = 100 \text{ centimetres (cm)} = 1000 \text{ millimetres (mm)}$$
$$1 \text{ centimetre} = 10 \text{ millimetres}$$

For day-to-day work in laboratories metre and half-metre rules are used — graduated in centimetres and millimetres. For more accurate measurement vernier calipers or a micrometer screw gauge may be used. Details of both these instruments will be found in any standard textbook. A metre rule is accurate to the nearest millimetre, calipers to the nearest 0.1 mm and a micrometer gauge to 0.01 mm.

## 1.2 Mass

The mass of a body measures the quantity of matter it contains. The unit of mass is the kilogram. A body will have the same mass in all parts of the universe.

$$1 \text{ kilogram (kg)} = 1000 \text{ grams (g)}$$
$$1 \text{ gram} = 1000 \text{ milligrams (mg)}$$

## 1.3 Area and volume

The area of a surface is measured in units of metres times metres ($m^2$) or $cm^2$ or $mm^2$. The volume of a substance is usually expressed in units of $m^3$ or $cm^3$ or $mm^3$.

The volume of a liquid is often measured in litres or millilitres.

$$1 \text{ litre (l)} = 1000 \text{ millilitres (ml)}$$
$$\text{but} \quad 1 \text{ millilitre} = 1 \text{ cm}^3 \text{ approximately}$$
$$\text{therefore} \quad 1 \text{ litre} = 1000 \text{ cm}^3 \text{ approximately}$$

In Physics the measuring cylinder is most commonly used for volume measurements of liquids. When reading the value it is important to look at the bottom of the curved liquid surface (meniscus).

## 1.4 Time

The scientific unit of time is the second (s) which is $\frac{1}{24 \times 60 \times 60}$ part of the time the earth takes to perform one revolution on its axis.

In the laboratory, time is normally measured using a stopclock or stopwatch. In some cases more accuracy is required, as, for example, when measuring the acceleration of a trolley moving on a ramp. A tickertape vibrator can then be used; see Unit 2.1 on speed, velocity and acceleration. The use of a centisecond or millisecond timer is explained in Unit 2.8. The sensitivities of these four instruments are as follows:

| | |
|---|---|
| stopwatch or stopclock | 1/10 s |
| tickertape vibrator | 1/50 s |
| centisecond timer | 1/100 s |
| millisecond timer | 1/1000 s |

In some experiments it may be appropriate to use a stroboscope to measure time. A description of the hand-operated stroboscope and its use is given at the end of Unit 14.2. Similarly the flashing stroboscope and its use are described in Units 2.8 and 2.9.

# 2 SPEED, VELOCITY AND ACCELERATION

## 2.1 Tickertape vibrator or 'ticker-timer'

The motion of an object moving in a laboratory can be studied with the aid of a tickertape, on which equal intervals of time are marked by a dot. The dots are printed by a 'ticker-timer' (Fig. 2.1), which consists of a flexible strip of soft iron $A$, clamped at one end $B$, and passing over one pole of a strong electromagnet $M$. A small alternating voltage from the mains (50 Hz) is connected to the electromagnet through a diode $D$, which only allows the resulting current to flow in one direction. Fifty times a second the current through the electromagnet grows and then dies away. Each time it grows the iron strip is attracted downwards. A stud or 'hammer' beneath the strip strikes a carbon disc below it, thus marking a tape $C$ running beneath the carbon. Each time the current dies away the strip is released. Thus 50 dots are made every second at equal intervals of $\frac{1}{50}$ second.

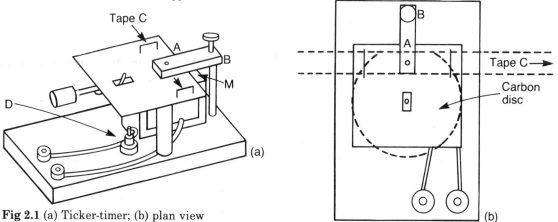

**Fig 2.1** (a) Ticker-timer; (b) plan view

## 2.2 Motion

Figure 2.2 shows tickertape pulled through the timer by a pupil who changed his speed as he went and on one occasion collided with another pupil.

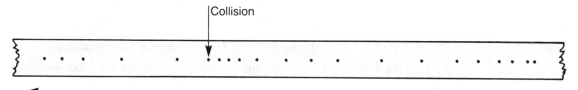

**Fig 2.2** Tickertape studies of motion

To show how the distance $s$ travelled by an object towing a length of tape varies with time $t$, the tape is cut into consecutive lengths containing equal numbers of dots (e.g. five). These lengths represent distances travelled in equal intervals of time (0.1 s), and so measure the speed of the object.

If they are arranged side by side (Fig. 2.3) a histogram of speed against time is obtained.

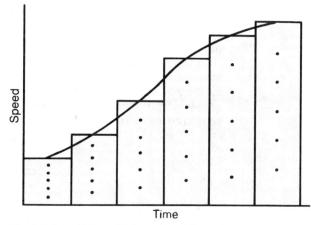

**Fig 2.3** Speed–time histogram and graph

In fact the speed does not change suddenly as indicated by the histogram. Each tape length is a measure of the average speed during $\frac{1}{10}$ second, whereas the speed is really changing all the time. A line drawn through the mid-points of the tape tops is a more accurate indication of the speed of the object at any moment. This line is a graph.

## 2.3 Speed

Speed is defined as the distance moved in one second.

$$\text{Average speed} = \frac{\text{distance moved}}{\text{time taken}} \text{ (m/s)}$$

Average speed may be obtained by measuring the distance travelled by an object while a certain number of dots (e.g. five) are made on the tape it is towing. For example, the average speed during each five-dot length of tape in Fig. 2.3 is calculated by dividing the length of each tape by 0.1 s. It is not, of course, necessary to cut up the tape in order to do this. The average speed can be calculated over any number of dots on any length of tape, providing the correct time interval is used in the calculation. If the dots on a measured length of tape are evenly spaced (i.e., the object is travelling with constant speed), then the speed calculated is the actual speed during that time interval.

## 2.4 Velocity

The velocity of a body measures its speed and the direction in which it travels.

$$\text{Velocity} = \frac{\text{distance moved in a particular direction}}{\text{time taken}}$$

(m/s in a particular direction — north for example)

Uniform velocity means that both the speed and the direction remain constant, as shown in Fig. 2.4(b).

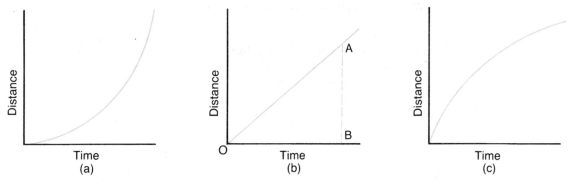

**Fig 2.4** Variation of distance moved in a straight line with time: (a) increasing velocity; (b) uniform velocity; (c) decreasing velocity

In Fig. 2.4(b) $AB$ represents the distance travelled in the time represented by $OB$, thus

$$\text{Velocity} = \frac{AB}{OB}; \text{ this is called the slope or gradient of the graph}$$

In Fig. 2.4(a) and (c) the gradients vary. Thus they represent bodies whose velocities have different values at different times.

## 2.5 Acceleration

If the velocity of a body is changing, the body is said to be accelerating. Acceleration is defined as the change in velocity per second.

$$\text{Acceleration} = \frac{\text{change in velocity}}{\text{time taken for this change}}$$

For example, suppose a car travelling along a straight road increases its speed from 10 m/s to 20 m/s in five seconds.

$$\text{Change in velocity} = (20 - 10) \text{ m/s} = 10 \text{ m/s}$$

$$\text{time taken for this change} = 5 \text{ s}$$

Hence using the formula above:

$$\text{acceleration} = 10 \text{ m/s in } 5 \text{ s} = 2 \text{ m/s in } 1 \text{ s}$$

$$\text{thus acceleration} = 2 \text{ m/s}^2$$

In this example the acceleration resulted from a change in the magnitude of velocity (speed). However, velocity can change in either magnitude or direction (circular motion). A change in either means the body is accelerating.

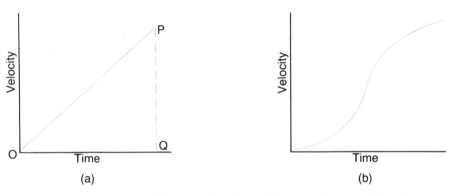

**Fig 2.5** Acceleration: (a) uniform acceleration; (b) non-uniform acceleration

In Fig. 2.5(a) the velocity is increasing with time at a steady rate, and the acceleration is said to be uniform. $PQ$ represents the change in velocity in time $OQ$.

$$\text{Acceleration} = \frac{PQ}{OQ}; \text{ the gradient of the velocity–time graph}$$

In Fig. 2.5(b) the acceleration is not constant. The acceleration at any instant is found by calculating the gradient of the graph at that time.

## 2.6 Uniformly accelerated motion

The body whose motion is represented by Fig. 2.6 is moving with a velocity $u$ when timing starts. It has a **uniform acceleration** of $a$ m/s² which means that each second its velocity increases by $a$, and after $t$ (seconds) its velocity will have increased by $at$. Hence at the end of this time its velocity $\mathbf{v = u + at}$.

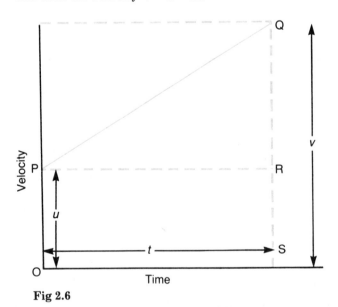

**Fig 2.6**

Alternatively,

$$\text{Acceleration } a = \frac{\text{change in velocity}}{\text{time taken for this change}}$$

$$= \frac{v - u}{t}$$

therefore $\quad v - u = at$

and $\quad \mathbf{v = u + at} \quad (2.1)$

## 2.7 Distance travelled

The average velocity of the body whose motion is shown in Fig. 2.6 is equal to half the sum of its initial velocity $u$, and final velocity $v$.

$$\text{Average velocity} = \frac{u+v}{2} \text{ (this only applies to a body accelerating uniformly)}$$

The distance $s$ moved can be found using the equation:

$$\text{Velocity} = \frac{\text{distance moved}}{\text{time taken}}$$

hence    distance moved = average velocity × time taken

$$s = \frac{(u+v)}{2} \times t \tag{2.2}$$

but    $v = u + at$ (2.1)

Therefore    $s = \dfrac{(u + u + at)t}{2}$

or    $\boldsymbol{s = ut + \tfrac{1}{2}at^2}$ (2.3)

In Fig. 2.6, $ut$ is the area of the rectangle $OPRS$. The area of the triangle $PQR$ is

$$PQR = \tfrac{1}{2} \times \text{base} \times \text{height}$$
$$= \tfrac{1}{2} \times t \times (v - u)$$

but    $v - u = at$   from equation (2.1)

Thus    area $PQR = \tfrac{1}{2}at^2$

therefore    area of $OPQS$ = (area of rectangle $OPRS$) + (area of triangle $PQR$)

Hence    area of $OPQS = ut + \tfrac{1}{2}at^2$

but    $s = ut + \tfrac{1}{2}at^2$ (2.3)

The area under the velocity–time graph in Fig. 2.6 is therefore equal to the distance travelled by the body. This is true for all such graphs, even when the acceleration is non-uniform.

A third equation for uniformly accelerated motion may be obtained by eliminating time $t$ between equations (2.1) and (2.3). The resulting equation is

$$\boldsymbol{v^2 = u^2 + 2as} \tag{2.4}$$

## 2.8 Experiments to show uniformly accelerated motion

### A TROLLEY MOVING DOWN A RAMP

One end of a ramp is raised several inches above a bench; the other end rests on the bench. A length of tickertape is fixed to the rear of a trolley, and the trolley allowed to run freely down the sloping ramp. As the tickertape passes through a ticker-timer dots are made on it at intervals of $\tfrac{1}{50}$ second. The tape is then cut into five-dot lengths and these are placed side by side to form a histogram (Fig. 2.7). Since the gradient is constant the acceleration must be uniform.

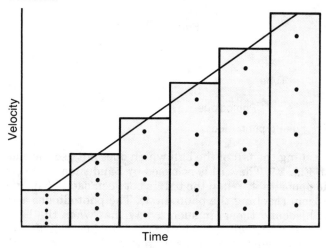

**Fig 2.7** Uniformly accelerated motion

# A BODY FALLING FREELY UNDER GRAVITY

In the last experiment the acceleration of the trolley was fairly small. A body falling freely under gravity has a much larger acceleration, and therefore the time of fall which has to be measured is small. It is usual to use a centisecond or millisecond timer for this purpose.

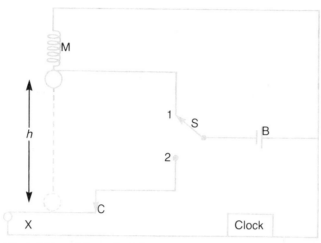

**Fig 2.8** Free fall method for g

The apparatus used for this experiment is shown in Fig. 2.8. It consists of an electromagnet $M$ connected to a battery $B$ via a switch $S$. The current passing through the electromagnet provides a magnetic field which holds a metal ball in place. When the switch is moved to position 2, the current stops and the ball begins falling. At the same time the circuit to the clock (centisecond or millisecond timer) is completed and it starts counting. After falling through a height $h$, the ball hits a hinged plate $X$, thus breaking the contact at $C$ and stopping the clock. The time $t$ is read to an accuracy of $\frac{1}{100}$ or $\frac{1}{1000}$ second. The distance $h$ is measured as accurately as possible; the value of the acceleration of gravity is then calculated using equation (2.3).

$$s = ut + \tfrac{1}{2}at^2$$

Here $s = h, u = 0, a = g$

thus $h = \tfrac{1}{2}gt^2$

hence $g = \dfrac{2h}{t^2}$

It is usual to repeat the experiment once or more as a check. The value of the acceleration due to gravity is approximately 9.8 m/s² at any point on the earth's surface. It has a much lower value on the surface of the moon.

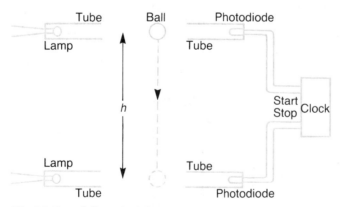

**Fig 2.9** Free fall method for g using photodiodes

An alternative way of timing the fall of the ball which does not use the electromagnet or the hinged plate is shown in Fig. 2.9. The ball is released by hand such that it falls between each lamp and the photodiode opposite it. When the ball is between each lamp and its photodiode it prevents light from the lamp reaching the photodiode. The photodiodes are connected to the clock (centisecond or millisecond timer) in such a way that when the ball passes the upper diode the clock is made to start and when the ball passes the lower diode the clock is stopped. The distance $h$ between the two lamps or photodiodes is measured with a rule and the acceleration due to gravity $g$ is calculated in the same way as above.

In a third method for doing this experiment a stroboscope is used. It may be either a hand-operated (Unit 14.2) or a flashing one. A flashing stroboscope emits short flashes of light at regular intervals. The time between each flash can be changed by means of a calibrated dial. For this experiment a suitable time interval between successive flashes is 0.05 s (20 flashes per second).

The stroboscope is arranged so that the flashes from it illuminate a white ball as it falls in a dark room. A millimetre scale is placed alongside the path of the falling ball. As the ball is released the shutter of a camera is opened and this is closed when the ball stops falling. The film in the camera thus records a series of images of the ball alongside the scale at known equal time intervals. The distance $h$ the ball falls during a known time $t$ is measured from the picture and the acceleration due to gravity $g$ calculated as in the previous methods.

In each of these alternative methods measurements can be made over different heights $h$. Whatever value of $h$ is used the value of $g$ is about the same; that is, $g$ is found to be constant.

## 2.9 Projectiles

So far we have only considered objects travelling in a straight line. In this unit we shall study objects which are being accelerated by the force due to gravity and at the same time are moving horizontally at a steady speed. One example of such motion is the path of a ball which is projected horizontally over the edge of a table. This ball continues to move with the same horizontal speed as it had just as it left the edge of the table (if we ignore the small effect of air resistance). In addition the ball falls under the influence of gravity.

It can be shown that the ball takes the same time to reach the ground as another ball dropped from the same height at the moment the first ball leaves the edge of the table. This shows that the horizontal motion of the first ball in no way affects its vertical acceleration; that is, it falls in exactly the same way as it would if it were not moving sideways. The horizontal and vertical motions of the ball can be treated entirely separately.

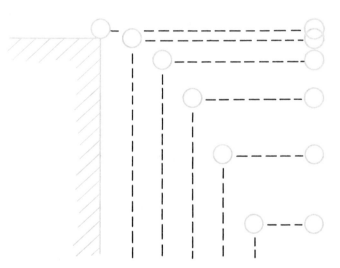

**Fig 2.10** Projectile motion

Figure 2.10 represents the motions of the two balls already mentioned. The paths of the two have been photographed at regular intervals. It can be seen that in equal time intervals both balls fall an increasing but equal distance (they undergo the same vertical acceleration). In the same equal time intervals the first ball moves equal distances horizontally. This motion may also be demonstrated using the pulsed water drops experiment. Another example of projectile motion is that of a ball thrown from one person to another.

## 2.10 Summary

1 A tickertape vibrator may be used for measuring speed, velocity or acceleration.

2 **Average speed** = $\dfrac{\text{distance moved}}{\text{time taken}}$ (m/s)

3 **Average velocity** = $\dfrac{\text{distance moved in a particular direction}}{\text{time taken}}$

(m/s in a particular direction — north for example)

Velocity is a vector quantity. It may increase, decrease or remain constant.

4 Acceleration = $\dfrac{\text{change in velocity}}{\text{time taken for this change}}$ (m/s$^2$)

Acceleration is a vector quantity. It may increase, decrease or remain constant — uniform acceleration.

5 For uniform acceleration, the following equations may be used.

$$s = \dfrac{(u+v)}{2} t$$

$$v = u + at$$

$$s = ut + \tfrac{1}{2}at^2$$

$$v^2 = u^2 + 2as$$

where $u$ is the initial velocity, i.e. at time $t = 0$
$v$ is the velocity at time $t$
$s$ is the distance travelled
$a$ is the uniform acceleration

6 In projectile motion the horizontal and vertical motions may be treated separately. The horizontal velocity is constant. The vertical acceleration is also constant and equal to that produced by gravity.

# 3 FORCE, MOMENTUM, WORK, ENERGY AND POWER

## 3.1 Newton's Laws of Motion

The majority of the work discussed in this unit can be summarised in Newton's Laws of Motion:

1 **Every body continues in a state of rest or uniform motion in a straight line unless acted on by a force.**
2 **When a force acts on a body the rate of change of momentum of the body is proportional to the force and the momentum changes in the direction in which the force acts.**
3 **To every action there is an equal and opposite reaction.**

The word **force** denotes a push or pull. When a body at rest is acted on by a force it tends to move; if a force acts on a body already in motion it will change its velocity, either by altering its speed or its direction, or both.

If a body has no force acting on it, the body will remain at rest or it will continue to move with a steady velocity.

Force is that which changes, or tends to change, a body's state of rest or uniform motion in a straight line.

A spring balance marked in newtons is suitable for measuring force.

The balance consists of a spring whose extension is proportional to the force applied to it.

The spring is contained in a case, and is calibrated by applying known forces to it, usually in the form of weights (Fig. 3.1).

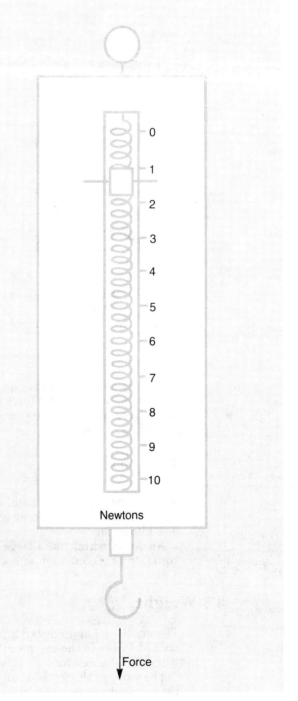

Fig 3.1

## 3.2 The acceleration produced by a force

The relationship between a force and the acceleration it produces can be investigated using a trolley, ramp and ticker-timer. The apparatus is the same as the one described in the last unit for showing uniformly accelerated motion. The ramp has to be raised sufficiently at one end to compensate for friction in the trolley and ticker-timer. The slope should be such that when the trolley with tape in place, is nudged, it moves slowly down the ramp at constant speed.

A tape is then obtained by towing the trolley down the ramp with a constant force. This can be provided by either an elastic band stretched a known amount, or by weights hung over a pulley as shown in Fig. 3.2. The experiment is repeated using twice the original force and again using three times the force.

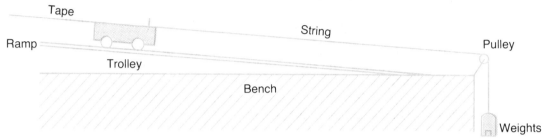

**Fig 3.2**

Histograms are made from the tapes in the usual way. A typical set of results is shown in Fig. 3.3.

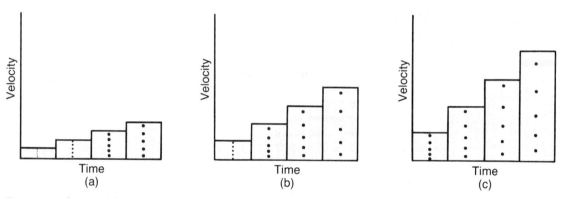

**Fig 3.3** (a) Original force; (b) twice the force; (c) three times the force

It is clear that if the force doubles, the acceleration doubles, etc., that is

$$\frac{F}{a} \text{ is constant, provided the mass is constant} \qquad (3.1)$$

The same apparatus may be used to show how the acceleration $a$ depends on the mass $m$ of a trolley, when a constant force is used. The force is applied in turn to one, two and three trolleys of similar mass stacked on each other. It is found that if the mass is doubled the acceleration is halved, etc., that is

$$m \times a = \text{a constant} \qquad (3.2)$$

The relation

$$\boldsymbol{F = ma} \qquad (3.3)$$

will be seen to include both statements (3.1) and (3.2). The unit of force is the **newton (N)** which is defined as the force which gives an acceleration of one metre per second$^2$ to a mass of one kilogram.

An object which has a large mass requires a large force to accelerate it as can be noted from equation (3.3). Such an object is said to have large inertia; it is difficult to move.

## 3.3 Weight

The acceleration produced by the earth when a body is falling freely is written as $g$. The force on a body due to the earth's attraction is found by using equation (3.3) which becomes $F = mg$, where $F$ is in newtons.

The value of the force acting on a body due to the earth's attraction is $mg$, or its weight. This force acts downwards, that is towards the centre of the earth.

The strength of the moon's attraction is about one seventh of the value of the earth's attraction. A mass $m$ on the moon's surface would therefore experience a force of only one seventh the value it experiences on the earth's surface. Its weight on the moon would be about one seventh of its weight on the earth.

## 3.4 Motion in a circle

When a body is moving in a circle its direction of motion, and hence its velocity, is continually changing. A force is needed to achieve this.

If you attach a mass to one end of a length of string, hold the other end and whirl it steadily round in a horizontal circle, you can feel that a force is required to make the mass move in a circular path. The force acts along the string and increases if the mass is whirled faster.

If any object is to move in a circle a force has to be continually applied to the object at right angles to its direction of motion at any instant; that is towards the centre of the circle. This is called the centripetal force. It does not alter the speed of the object, but it does alter its direction of travel and hence its velocity. The force produces an acceleration towards the centre of the circle.

Consider an object of mass $m$ moving with a constant speed $v$ in a circle of radius $r$. By calculating the rate of change of *velocity*, as the object continually changes direction, the acceleration $a$ can be shown to be given by the equation:

$$a = \frac{v^2}{r}$$

but $\quad F = ma$

thus $\quad F = \dfrac{mv^2}{r} \quad\quad\quad (3.4)$

The relationship shown in equation (3.4) can be investigated using the apparatus illustrated in Fig. 3.4. It consists of a mass $S$, such as a rubber stopper, at the end of a long piece of string or cord $C$, which passes through a glass tube with a rubber grip. The ends of the glass tube are polished so that friction is reduced and there is little danger of the string fraying. The force $F$ is provided by a number of steel washers $W$ of equal weight attached to the end of the string.

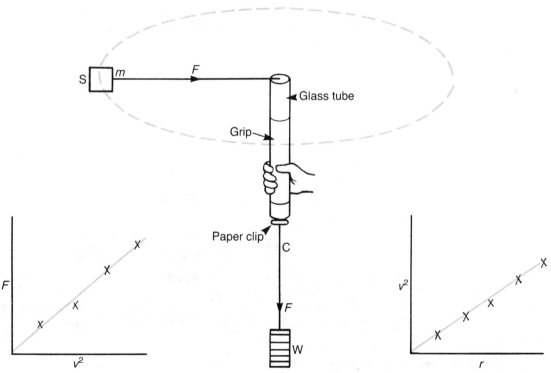

**Fig 3.4** Centripetal force

The mass $S$ is swung in a horizontal circle at a steady speed, with as little movement of the hand as possible. Care is taken that the paper clip rotates freely just below the grip. This ensures that the radius remains constant. The time is noted for a given number of revolutions: 20 for example. Then

$$v = \frac{2\pi r n}{T}$$

where $n$ is the number of revolutions counted and $T$ is the time noted.

The number of washers is then increased and the experiment repeated using the same radius. A series of results is obtained in this way. A further series is obtained using a constant force, but changing the radius.

The graphs shown in Fig. 3.4, are plotted. They show that:

1. $\dfrac{F}{v^2}$ is constant for a constant radius

2. $\dfrac{v^2}{r}$ is constant for a constant force

The equation $\boldsymbol{F = \dfrac{mv^2}{r}}$ includes both these statements.

If the string had broken at any time the mass $S$ would have flown off at a tangent to the circle; that is, on the removal of the force, the mass would have continued in a straight line with a steady speed.

A car or train going round a corner are everyday examples of motion in a circle. Friction between the tyres and the road provides the centripetal force in the case of a car cornering. If the corner is too sharp ($r$ very small), the speed too high, or the road wet, the frictional force may not be large enough to keep the car moving in a circle, and it will slide off the road. The outer rail provides the centripetal force in the case of a train.

Another example of centripetal force is the gravitational force of the sun on the planets, including the earth.

## 3.5 Momentum

A lorry which is fully laden requires a larger force to set it in motion than a similar lorry which is empty. Likewise more powerful brakes are required to stop a heavy goods vehicle than a family car moving with the same speed. The heavier vehicle is said to have more **momentum** than the lighter one. Momentum is a measure of how difficult it is to alter a body's motion; it is more basic even than velocity.

The momentum of a body is defined as the product of its mass and its velocity.

**Momentum = mass × velocity**

By studying the linear motion of a trolley down a ramp it has been shown experimentally that the following equation is valid:

$$F = ma$$

but $\quad a = \dfrac{v - u}{t}$

Thus $\quad F = \dfrac{m(v - u)}{t}$

so $\quad F = \dfrac{mv - mu}{t} \quad$ (3.5)

or $\quad \boldsymbol{Ft = mv - mu} \quad$ (3.6)

Equation (3.5) can be written thus:

$$\text{Force} = \frac{\text{change of momentum}}{\text{time taken for this change}}$$

All the equations just discussed concern momentum change and hence, by the definition of momentum, velocity change. They are therefore valid for changes in the direction of motion of a body as well as for changes in its speed. They apply to circular motion as well as linear, and are summed up in Newton's second law of motion (see Unit 3.1).

## 3.6 Action and reaction

**When two bodies collide the force one exerts on the other is equal in size but opposite in direction to the force the second one exerts on the first.** As the time of contact is the same for each both experience the same change in momentum, but in the opposite direction. The total momentum of the two is thus unaltered by the collision. This is known as **the principle of conservation of linear momentum**. The principle always holds providing no forces, other than those due to the collision, act on the bodies.

This principle can be verified experimentally, for a simple case, using two trolleys and a ticker-timer, on a friction-compensated ramp. Fig. 3.5(a) shows the arrangement.

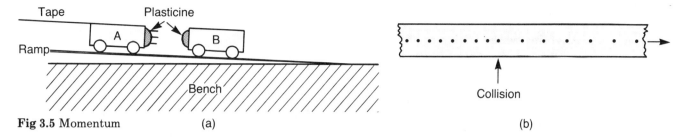

**Fig 3.5** Momentum    (a)    (b)

Some plasticine is placed on the surface of the two trolleys which will come into contact. Drawing pins are embedded in the plasticine on the front of trolley A, and tickertape is attached to the rear of this trolley. Trolley A is given a push so that it travels with a uniform velocity and collides with trolley B. The trolleys stick together and move off with a common velocity. A typical tape is shown in Fig. 3.5(b). The spacing of the dots changes abruptly on collision. The velocity of A before the collision and the combined velocity of the two trolleys after the collision are found from the spacing of the dots. In the experiment the tape is **NOT** being used to show acceleration. The results show that:

$$\text{Total momentum before collision} = \text{total momentum after collision} \tag{3.7}$$

Alternatively

$$m_1 u_1 + m_2 u_2 = m_1 v_1 + m_2 v_2 \tag{3.8}$$

where $m$, $u$ and $v$ represent the masses and velocities of the colliding bodies. In some collisions one of the velocities may be zero or two velocities may have the same value. For example in the experiment just described $u_2 = 0$ and $v_1 = v_2$, as the bodies stick together after collision. Equation (3.8) then becomes simpler.

Suppose that in the above experiment $m_1 = 1$ kg, $m_2 = 3$ kg and $u_1 = 1$ m/s. Then substituting in equation (3.8) we obtain:

$$1 = v_1 + 3v_2$$

but $\quad v_1 = v_2$

thus $\quad 1 = 4v_1$

hence $\quad v_1 = 0.25 \text{ m/s} = v_2$

This is the value we would record from the tape after the collision, thus verifying **the principle of the conservation of momentum**.

The principle of conservation of momentum holds in the case of an explosion as well as a collision. A rocket relies for its propulsion on the fact that it gains a forward momentum equal in size to the momentum of the gases it expels backwards. It will gain forward momentum, and hence increase speed, for as long as it continues to expel these gases.

The principle may be used to estimate the speed of a pellet fired from an air rifle. The experiment can be done in the laboratory using the apparatus shown in Fig. 3.6. The pellet is fired into a piece of plasticine embedded in a block of wood mounted on a suitable vehicle. The vehicle is stationary on a track which has been compensated for friction. The speed of the vehicle after the collision is found by timing its motion over a measured length of track with a stopwatch. The mass of one pellet is determined by weighing 10 or 20 similar pellets.

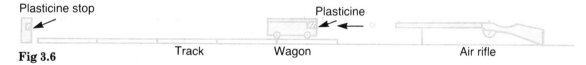

**Fig 3.6**    Track    Wagon    Air rifle

The initial speed of the pellet is calculated using equation (3.8) above. Again $u_2 = 0$ and $v_1 = v_2$. All the other quantities in the equation can be measured, except $u_1$, which is calculated from it. The value is likely to be about 100 m/s.

## 3.7 Work

The term 'work' is associated with movement. If a railway engine pulls a train along a track with a steady force the work done by the engine depends on the size of the force it provides and the distance it pulls the train with this force. Work is calculated by definition from the relation:

**Work done = force × distance moved in the direction of the force**

The unit of work is the joule (J). One joule is the work done when a force of one newton moves through a distance of one metre in the direction of the force.

No work is done in circular motion as the force is at right angles to the distance.

## 3.8 Energy

Anything which is able to do work is said to possess energy.

Energy is the capacity to do work. All forms of energy are measured in joules.

The world we live in provides energy in many different forms, of which chemical energy is perhaps the most important. The use of chemical energy from coal and oil has been a major factor in the development of our civilization. The presence of electricity, light and heat as forms of energy in our homes is something we take for granted. All these are produced from the chemical energy released from coal or oil or from nuclear energy.

Energy can neither be created nor destroyed, though it may be changed from one form to another. This is a statement of **the law of conservation of energy**.

Although energy may change from one form to another, the second form may not be measurable or useful. For example this is true of the heat produced when a frictional force is overcome.

## 3.9 Potential energy

If a body is to be raised from a bench, an upward vertical force equal to the weight of the body must be provided. Suppose the force applied ($mg$) raises the body a vertical distance $h$.

Work done = force × distance moved in the direction of the force

Thus  Work done = $mg \times h =$ **$mgh$**

Once the body has been raised it is said to have increased its **potential energy** by this amount. All the work done has been used to increase the potential energy of the body.

Potential energy is the energy a body has by reason of its position.

## 3.10 Kinetic energy

If the mass is now allowed to fall it will steadily lose the potential energy it has gained. By the time it has fallen a distance $h$ it will have lost all the potential energy it previously gained. As it falls its velocity increases and it is said to possess an increasing amount of **kinetic energy**. The kinetic energy it has at any instant will equal the potential energy it has lost.

Kinetic energy is the energy a body has by reason of its motion.

As the mass falls it is accelerated by the force of the earth's attraction on it. Suppose it falls a distance $h$, in time $t$, as its velocity increases uniformly from zero to $v$.

$$\text{Force} = \frac{\text{change in momentum}}{\text{time taken}}$$

Thus $$F = \frac{mv - 0}{t}$$

and  distance = average velocity × time

that is $$h = \frac{v}{2} \times t$$

Work done = force × distance moved in the direction of force

$$= \frac{mv}{t} \times h$$

$$= \frac{mv}{t} \times \frac{vt}{2}$$

$$= \tfrac{1}{2}mv^2$$

The work done by the gravitational force results in an increase of kinetic energy of $\tfrac{1}{2}mv^2$, and a loss of potential energy of $mgh$.

Suppose a body, mass 5 kg, falls through a vertical distance of 5 m near the earth's surface where its acceleration due to gravity ($g$) is 10 m/s².

Potential energy lost = $mgh = 5 \times 10 \times 5 = 250$ J

Kinetic energy gained = $\tfrac{1}{2}mv^2 = 250$ J

Thus  $\tfrac{1}{2} \times 5 \times v^2 = 250$ J

$v^2 = 100$

and  $v = 10$ m/s

## 3.11 Power

Power is defined as the work done per second, or the amount of energy transformed per second.

$$\text{Average power} = \frac{\text{work done}}{\text{time taken}} = \frac{\text{energy change}}{\text{time taken}}$$

It is measured in units of joules per second (J/s). One joule per second is called a watt (W).

$$1 \text{ kilowatt (kW)} = 1000 \text{ watts}$$

A rough estimate of a pupil's power can be made by asking him to walk or run up a staircase. If the height of the staircase is measured in metres, the pupil's weight calculated in newtons, and the time he takes recorded on a stopwatch, the power can be calculated. The result will normally be about 200 W if walking or 500 W if running.

## 3.12 Scalar and vector quantities

A **scalar quantity** is one which has only magnitude (size), such as money and number of apples. A **vector quantity** is one which has both magnitude and direction, such as velocity, force and momentum.

Vectors can be represented by straight lines drawn to scale. If a number of forces all act in the same straight line, their resultant is determined by addition or subtraction.

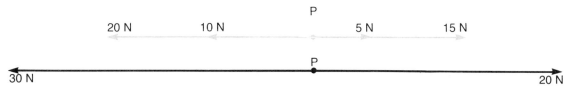

**Fig 3.7**

Thus if forces of 20, 15, 10 and 5 newtons all act at a point $P$, as shown in Fig. 3.7, we have:

$$\text{total force acting towards the left} = 10 + 20 = 30 \text{ N}$$
$$\text{total force acting towards the right} = 15 + 5 = 20 \text{ N}$$
$$\text{resultant force} = 30 - 20 = 10 \text{ N acting to the left}$$

## 3.13 Summary

1. When a body is in motion it obeys Newton's laws of motion.
2. The acceleration $a$ produced when a mass $m$ experiences a force $F$ is given by

$$a = \frac{F}{m} \text{ or } F = ma$$

3. When a body mass $m$ moves at constant speed in a circle of radius $m$, the centripetal force $F$ is given by

$$F = \frac{mv^2}{r}$$

4. **Momentum = mass × velocity ($mv$)** and is measured in units of kg m/s.
5. The law of conservation of momentum states that when two bodies collide the total momentum in a given direction after collision is equal to the total momentum in the same direction before collision.
6. The change of momentum which occurs when a force acts on a body for a certain time is given by

$$\text{Force} = \frac{\text{change in momentum}}{\text{time taken for this change}}$$

Force is measured in newtons (N).

7. **Work done (or energy transformed) = force × distance moved in the direction of the force.**
8. The change in potential energy (or energy of position) when a mass $m$ is moved through a vertical height $h$ is given by the equation

$$\text{Change in potential energy} = mgh$$

where $g$ is the gravitational field strength.

9 The kinetic energy (or energy of motion) of a mass $m$ moving at a velocity $v$ is given by

$$\text{Kinetic energy} = \tfrac{1}{2}mv^2$$

10 Power is the work done per second or the rate at which energy is transformed

$$\text{Average power} = \frac{\text{work done}}{\text{time taken}} = \frac{\text{energy change}}{\text{time taken}}$$

11 Energy is measured in joules (J) and power in watts (W). One watt is one joule per second.

12 A **scalar** quantity has only size. A **vector** quantity has size and direction.

# 4 TURNING FORCES AND MACHINES

When we open a door, turn on a tap or use a spanner, we exert a **turning force**. Two factors determine the size of the turning effect: the magnitude of the force and the distance of the line of action of the force from the pivot or fulcrum. A large turning effect can be produced with a small force provided the distance from the fulcrum is large. The size of the turning effect is called the **moment**.

The moment of a force about a point is the product of the force and the perpendicular distance of its line of action from the point.

**Moment of force = force × perpendicular distance from the pivot**

## 4.1 Experiment to study moments

A thin uniform strip of wood, for example, a metre rule, is balanced on a fulcrum. A weight is placed on the rule to one side of the fulcrum, a second weight added on the other side and its position carefully adjusted until balance is restored (Fig. 4.1). Within the limits of experimental error it will be found that:

$$w_1 \times d_1 = w_2 \times d_2$$

Balance can be restored using more than one weight, in which case:

$$w_1 \times d_1 = (w_2 \times d_2) + (w_3 \times d_3)$$

This principle can be extended if further weights are used. When a body is in equilibrium, the sum of the anti-clockwise moments about any point is equal to the sum of the clockwise moments about the same point. The force the fulcrum exerts on the body equals the sum of the turning forces on the body.

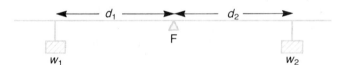

**Fig 4.1** Moments

Suppose, in Fig. 4.1 above, $w_1 = 300$ N, $w_2 = 100$ N, $d_1 = 0.4$ m and $d_2 = 0.5$ m, then we can calculate where a weight of 700 N must be placed to balance the rule.

$$w_1 \times d_1 = (w_2 \times d_2) + (w_3 \times d_3)$$

hence $\quad 300 \times 0.4 = (100 \times 0.5) + 700 d_3$

and $\quad d_3 = \dfrac{(300 \times 0.4) - (100 \times 0.5)}{700}$ m

$$= \dfrac{120 - 50}{700} = 0.1 \text{ m}$$

This weight must be placed 0.1 m to the right of the pivot.

## 4.2 Centre of mass

The weight of a body is defined as the force with which the earth attracts it. This says nothing about the point of application of this force. A body may be regarded as made up of a large number of tiny particles, each with the same mass. Each of these particles is pulled towards the earth with the same force. The earth's pull on the body thus consists of a large number of equal parallel forces. These can be replaced by a single force which acts through a point called the **centre of mass**.

The centre of mass of a body is defined as the point of application of the resultant force due to the earth's attraction on the body. Thus we may regard the centre of mass as the point at which the whole weight of the body acts.

## 4.3 Location of the centre of mass

The centre of mass of a long thin object such as a ruler may be found approximately by balancing it on a straight edge. The same method may also be used for a thin sheet (lamina). In this case it is necessary to balance it in two positions as shown in Fig. 4.2.

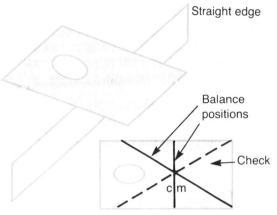
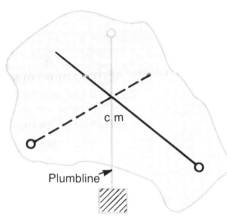

**Fig 4.2** Locating the centre of mass      **Fig 4.3**

One good way of finding the centre of mass of a lamina is to use a plumbline. Three small holes are made at well spaced intervals round the edge of the lamina, and the lamina and plumbline suspended from each in turn (Fig. 4.3).

The position of the plumbline is marked on the lamina and the point of intersection of these three lines gives the position of the centre of mass.

## 4.4 Stability

The position of a body's centre of mass affects its stability. For example, the centre of mass of a vehicle should be as low as possible, and its wheel base as wide as possible, if it is to be stable.

When a vehicle corners fast there is a tendency for it to tilt on the outer wheels. It will turn right over when the vertical line through the centre of mass falls outside these wheels. If the conditions stated above are satisfied the centre of mass has to rise a larger distance for this to happen. This requires more potential energy and it is less likely to occur.

## 4.5 Machines

A machine is any device by means of which a force (effort) applied at one point can be used to overcome a force (load) at some other point. Most machines, but not all, are designed so that the effort is less than the load. They can be regarded as force multipliers; the force that the effort exerts is multiplied by the machine to be equal to the force that the load exerts.

The **mechanical advantage** of any machine is defined as the ratio of the load to the effort:

$$\text{Mechanical advantage} = \frac{\text{load}}{\text{effort}}$$

For example, if a lever can be used to overcome a load of 500 N by the application of an effort of 100 N, it has a mechanical advantage of five.

The ratio of the distance moved by the effort to the distance moved by the load in the same time is called the **velocity ratio** of the machine.

$$\text{Velocity ratio} = \frac{\text{distance moved by the effort}}{\text{distance moved by the load in the same time}}$$

Both mechanical advantage and velocity ratio are dimensionless; that is, they have no units.

## 4.6 Work done by a machine: efficiency

The ratio of the useful work done by a machine to the total work put into it is called the **efficiency** of the machine. Usually the efficiency is expressed as a percentage.

$$\text{Efficiency} = \frac{\text{work output} \times 100\%}{\text{work input}}$$

Since work done = force × distance, it follows that:

$$\text{efficiency} = \frac{\text{load} \times \text{distance load moves} \times 100\%}{\text{effort} \times \text{distance effort moves}}$$

thus

$$\text{efficiency} = \frac{\text{mechanical advantage} \times 100\%}{\text{velocity ratio}}$$

In a perfect machine no work would be wasted and the efficiency would be 100%. It follows that the mechanical advantage and velocity ratio are then equal. In practice, work is wasted in overcoming friction and, in the case of pulleys, in raising the lower pulley block. The efficiency is then below 100%.

## 4.7 The lever

The lever is the simplest form of machine in common use. It can consist of any rigid body pivoted about a fulcrum. Levers are based on the principle of moments discussed earlier in this section. A force (effort) is applied at one point on the lever, and this overcomes a force called the load at some other point. Fig. 4.4 illustrates some simple machines based on the lever principle.

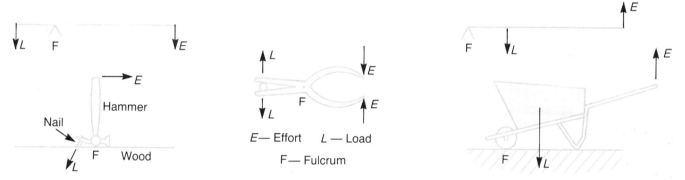

**Fig 4.4** Simple levers

Details of other machines such as the screw jack, inclined plane and gears may be found by reference to standard textbooks.

## 4.8 Pulleys

A pulley is a wheel with a grooved rim. Often one or more pulleys are mounted together to form a block. The use of pulleys is best illustrated by considering the block and tackle arrangement shown in Fig. 4.5.

For clarity the pulleys are shown on separate axles; in practice the two pulleys in the top block are mounted on a single axle as are those in the bottom block. In order to raise the load by one metre each of the four strings supporting the lower block must be shortened by 1 m. This is achieved by the effort being applied through a distance of 4 m.

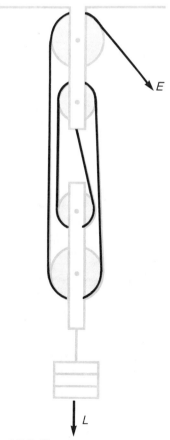

**Fig 4.5** Pulley system

## 4.9 Summary

1. The turning effect of a force about a pivot is called its moment.

    **Moment of a force = force × perpendicular distance from the pivot**

    It is measured in newton metres (Nm).

2. A machine is any device by which a force (effort) applied at one point can be used to overcome a force (load) at some other point. Most machines are designed so that the effort is less than the load.

    $$\text{Mechanical advantage} = \frac{\text{load}}{\text{effort}}$$

    The efficiency of a machine is given by

    $$\text{Efficiency} = \frac{\text{work output}}{\text{work input}} \times 100\%$$

3. Any lever or pulley system is a type of machine.

# 5 DENSITY

Equal volumes of different substances vary considerably in their mass. For instance aircraft are made chiefly from aluminium alloys which, volume for volume, have a mass half that of steel, but are just as strong. The 'lightness' or 'heaviness' of a material is referred to as **its density**.

$$\text{Density} = \frac{\text{mass}}{\text{volume}} \; (\text{kg/m}^3 \text{ or g/cm}^3)$$

## 5.1 Density measurements

### A REGULAR SOLID

The mass of the solid is found by weighing, either on a chemical balance, if great accuracy is required, or on a spring balance. If the latter is used the result must be converted to mass units; that is if the balance is calibrated in newtons, the value recorded must be divided by 9.81 to obtain the mass in kilograms.

The volume of the solid is obtained by length measurement, using a ruler, vernier callipers or a micrometer screw gauge, depending on the accuracy required. This method is applicable to cuboids, spheres, cylinders and cones amongst other regular shapes. The formula giving the volume of such shapes in terms of their linear dimensions can be obtained from textbooks. For example, metals are often in the form of turned cylinders whose volume can be calculated from the formula:

$$\text{Volume} = \pi r^2 h$$

where $r$ is the cylinder radius and $h$ is its height.

### AN IRREGULAR SOLID

The mass is found in the same way as for a regular solid. In order to find the volume it is necessary to partly fill a measuring cylinder with water. The reading is taken and the solid then lowered into the water on the end of a length of cotton, until it is completely immersed, and the new reading taken. The difference between the two readings gives the volume of the solid. This method cannot be used if the solid dissolves in water.

### A LIQUID

A measuring cylinder is first weighed empty using, for example, a top-pan balance. Some of the liquid to be tested is poured into the cylinder, and the cylinder reweighed. The difference between the two readings gives the mass. The volume of the liquid is obtained by direct reading of the measuring cylinder. If a more accurate value is required a specific gravity bottle can be used. Details may be obtained from a textbook.

### A GAS

Air is the most common gas. Its density may be found by pumping air into a large plastic container. The container, complete with an open tap, is first weighed; a lever arm balance is

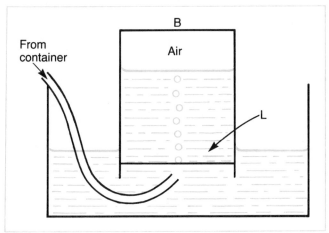

Fig 5.1

suitable for this purpose. Air is then pumped into the container until it is very hard, using a bicycle or foot pump and a one-way valve in the connecting tube. The tap is then closed and the container re-weighed. By means of tubing attached to the tap, the air is now released from the container and collected over water as shown in Fig. 5.1.

A transparent plastic box $B$ makes a suitable collecting container. A line $L$ is marked on it so that one litre is collected. The box is filled with water and inverted as shown. Air slowly bubbles through the tube, and the box is moved up or down until the outside and inside water levels are in line with the mark $L$. One litre of air at atmospheric pressure has now been collected. The tap is closed, the plastic box refilled with water and the bubbling process repeated. This is continued until no more air bubbles through. The final fraction of a litre collected has to be estimated.

The mass and volume of the extra air originally pumped into the container are now known and the density of air at atmospheric pressure may be calculated.

# 6 PRESSURE

The word **pressure** has a precise scientific meaning. It is defined as the force acting normally (perpendicularly) per unit area. For example, the pressure exerted on the ground by a body depends on the mass of the body and on the area of the body in contact with the ground. A boy wearing ice skates will exert a far greater pressure than if he were wearing shoes. The pressure exerted on the ground by a brick depends on which face is in contact with the ground. The weight of the brick and thus the force it exerts on the ground is about 22 N, whichever face is in contact (Fig. 6.1).

$$\text{Pressure} = \frac{\text{force}}{\text{area}}$$

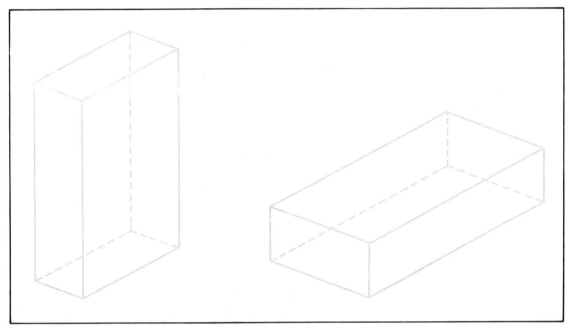

Fig 6.1

One does not have to use a very large force when using a needle. As the area of the point is very small, a relatively small force produces a large pressure and the needle pierces the material.

## 6.1 Pressure in a liquid or gas (a fluid)

The pressure in a fluid increases with depth. This may be shown by using a tall vessel full of water with side tubes fitted at various depths (Fig. 6.2).

The speed with which the water spurts out is greatest for the lowest jet, showing that pressure increases with depth. This demonstration also indicates that pressure acts in all directions in a fluid — not just vertically. The pressure responsible for these jets is acting horizontally.

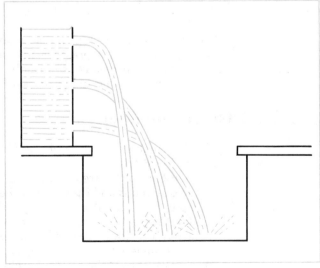

Fig 6.2

Suppose we consider a horizontal area $A$ at a depth $h$ below the surface of a fluid of density $d$. Standing on this area is a vertical column of liquid of volume $hA$, the mass of which is $hAd$. The weight of this mass is $hAdg$.

$$\text{Pressure} = \frac{\text{force (weight)}}{\text{area}} = \frac{hAdg}{A}$$
$$= hdg$$

The usual units are newtons per metre$^2$, often called pascals (Pa). The area does not appear in the final expression for pressure.

The property of liquids to transmit pressure to all parts is used in many appliances. Some car jacks consist of an oil-filled press used for lifting (Fig. 6.3).

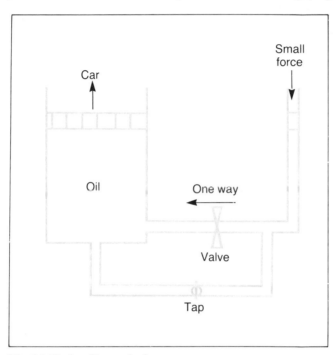

**Fig 6.3** Hydraulic car jack

The pressure exerted on one side of the press is transmitted through the liquid to the other side. Thus a small force applied over a small area on one side can result in a large force over a large area on the other side. A large mass, such as a car, may thus be raised by application of a small force.

Mechanical diggers and bulldozers use hydraulic principles to power the blade or shovel. Cars require a braking system which exerts the same pressure on the brake pads of all four wheels to reduce the risk of skidding. Each brake consists of two brake shoes which are pushed apart by hydraulic pressure in a cylinder and press on the brake drum. When the brakes are applied the increased pressure on the pedal is transmitted through oil to the cylinders in each wheel and the shoes applied.

## 6.2 Atmospheric pressure

On earth we are living under a large volume of air. Air has weight and as a result the atmosphere exerts a pressure not only on the earth's surface but on objects on the earth. **Atmospheric pressure** is normally expressed in newtons per metre$^2$ or Pascals. The average value over a long period is 100 000 Pa approximately.

## 6.3 The manometer

This instrument is used for measuring the pressure of a gas. It consists of a U-tube containing water (Fig. 6.4).

When both arms are open to the atmosphere, the same pressure is exerted on both water surfaces $A$ and $B$ and these are at the same level. In order to measure the pressure of a gas (for example, the laboratory gas supply) one arm of the manometer is connected by means of flexible tubing to the gas supply. When the gas supply is turned on it exerts a pressure on surface $A$; this level falls under the extra pressure and the level at $B$ rises, until the pressure at $C$, on the same level as $A$, becomes equal to the gas pressure. The excess pressure of the gas supply over that of the atmosphere is given by the pressure due to the water column $BC$, and is equal to $hdg$ (N/m$^2$).

The height $h$ is known as the head of water in the manometer and it is common to use this height as a measure of the excess pressure. If relatively high pressures are to be measured using the manometer it must be filled with a more dense liquid such as mercury.

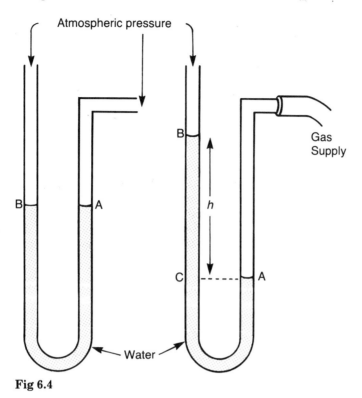

**Fig 6.4**

## 6.4 Simple barometer

The simple barometer uses the same principle as the manometer; the reservoir takes the place of one arm. As atmospheric pressure is to be measured it is necessary to have a vacuum above the mercury in the tube (Fig. 6.5).

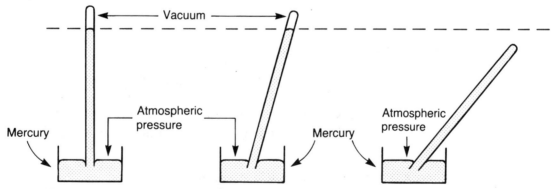

**Fig 6.5** Simple barometer

A simple barometer may be made by taking a thick-walled glass tube about one metre long and closed at one end, and filling it almost to the top with mercury. A finger is then placed securely over the end and the tube inverted several times. The large air bubble which travels up and down collects the small ones clinging to the side of the tube. More mercury is then added so that the tube is completely full. A finger is again placed over the open end, the tube inverted carefully, its open end placed under the surface of mercury in a reservoir or dish, and the finger removed. The mercury falls until the vertical difference in level between the two mercury surfaces is about 76 cm. Small changes in the level of the mercury in the tube occur from day to day due to small changes in atmospheric pressure.

## 6.5 Aneroid barometer

An aneroid barometer contains no liquid. The basis of the instrument is a flat cylindrical metal box which has been partially evacuated and sealed (Fig. 6.6).

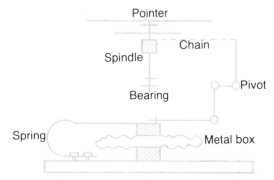

**Fig 6.6** Aneroid barometer

Increase in atmospheric pressure causes the box to cave in slightly; a decrease allows it to expand slightly. These movements of the box are magnified by a system of levers. The levers are connected to a fine chain wound round the spindle of a pointer. A hairspring keeps the chain taut. The pointer moves over a dial usually calibrated as shown in Fig. 6.7.

The dial of an aneroid barometer can be calibrated in feet for use as an altimeter in aircraft. Atmospheric pressure falls by about $\frac{1}{3}$ during the first 10 000 feet of ascent above the earth's surface.

**Fig 6.7**

## 6.6 Summary

1 The word 'pressure' has a precise meaning.

$$\text{Pressure} = \frac{\text{force}}{\text{area}}$$

It is measured in N/m² which is often called a Pascal (Pa).

2 The pressure a distance $h$ below the surface of a liquid which has a density $d$ is

$$\text{pressure} = hdg$$

where $g$ is the gravitational field strength.

3 A pressure applied at one point in a liquid is transmitted to all other points. The brake systems of cars and some car jacks use this principle.

4 The earth's atmosphere exerts a pressure at the earth's surface of about 100 000 Pa. It can be measured using a mercury or aneroid barometer.

# 7 FORCES BETWEEN MOLECULES IN SOLIDS

Before materials are used in the construction of machinery, bridges and buildings, tests are carried out to ensure that they are able to stand up to the stresses to which they are likely to be subjected. Brittle substances such as cast iron and masonry will support large forces of compression but break easily if stretching forces are applied. When stretching forces are likely to be significant, materials such as steel have to be used. The behaviour of a material under the influence of applied forces depends on the forces holding the molecules of the material together.

## 7.1 Elasticity

Some knowledge of the forces between molecules of a solid can be gained by adding weights to a spiral spring and investigating how it stretches. A spiral spring is suspended vertically from a rigid support and a small pointer attached to its lower end (Fig. 7.1(a)). The reading of the pointer against the scale is noted. Weights are then added in steps to the lower end of the spring and the reading of the pointer recorded after the addition of each weight. The weights are removed in similar steps and a second set of readings taken. For small loads the reading of the pointer should be the same for each set of readings. The average extension for each load is then plotted against the load (Fig. 7.1(b)).

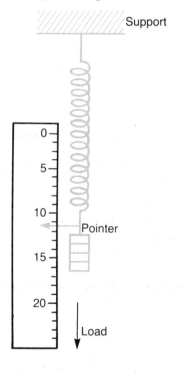
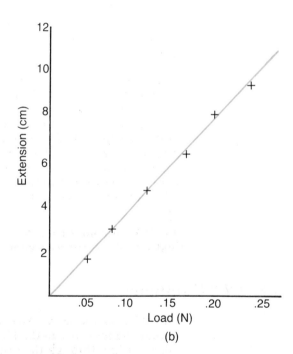

**Fig 7.1** (a)            (b)

A straight-line graph passing through the origin shows that the extension is directly proportional to the load in the range of loads used: that is, the load/extension ratio is constant. When a small weight is attached to the spring, the spring extends. When the weight is removed the spring returns to its original length. The property of regaining its original size or shape is called **elasticity** — thus putty is a very inelastic material whereas metals regain their original shape and are therefore elastic. In a metal the forces of attraction between the displaced molecules are sufficiently strong to restore the molecules to their original position.

If larger weights had been added to the spring a stage would have been reached when the spring would not have returned to its original shape on removing the weights. The spring is permanently stretched or deformed. Beyond a certain load the molecules do not return to their original positions when the load is removed. The extension at which this occurs is called the **elastic limit** of the spring. With greater loads the molecules are unable to keep their fixed positions in the metal.

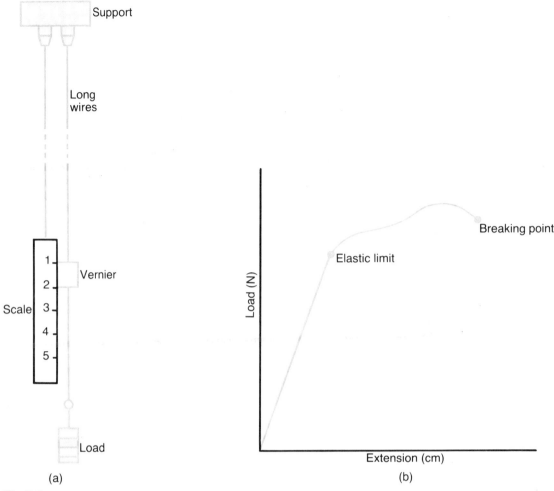

**Fig 7.2**

The forces between the molecules of a metal can be further investigated by stretching a length of straight wire. A length of wire is suspended vertically from a fixed support and weights added to its lower end, in a similar way to the procedure for a spring. However, the extension of the wire within its elastic region is very small, and if this region is to be studied a more accurate method of measuring extension has to be used. It is usual to use two wires, one carrying a vernier and the other a millimetre scale (Fig. 7.2(a)). The second wire is for comparison purposes and is not stretched. This comparison method of measuring extension eliminates errors due to thermal expansion and sag of the support.

The results from these two experiments may be summarised in what is known as Hooke's law:

**The deformation of a material is proportional to the force applied to it, provided the elastic limit is not exceeded.**

## 7.2 Summary

1 An object is said to behave elastically when equal increases in the force applied to it produce equal changes in length. That is, a graph of the force applied against the extension is a straight line through the origin. When this is so Hooke's law is said to be obeyed.
2 The elastic limit is the point beyond which the object no longer behaves elastically.

# 8 KINETIC THEORY OF MATTER

## 8.1 Atoms and molecules

In 1808 John Dalton produced experimental evidence to show that chemical compounds consist of molecules. A molecule is a group of atoms. There are about 90 different chemical elements occurring in nature, each of which has its own characteristic atom.

Some idea of the size of molecules can be gained from an experiment to measure the thickness of a very thin oil film. A tank with a large surface area is required for this experiment and the water which is placed in the tank must be very clean. Lycopodium or talcum powder is lightly sprinkled on the water surface. A small oil drop is then placed on the surface of the water near the centre of the tank. It can be transferred to the water using a loop of fine wire or a small syringe. Immediately the oil drop touches the water surface the powder is pushed back by the oil film to form a ring of clear water, the diameter of which is measured.

The oil film may be considered as a cylinder of radius $R$ and height $h$. Hence its volume $= \pi R^2 h$. The value of $h$ is calculated by equating the volume of the oil film to the volume of the oil drop placed on the water surface. On extremely clean water, olive oil is considered to spread until the film is one molecule thick. In this case $h$ is an estimate of the diameter of an oil molecule. The value obtained is about $2 \times 10^{-9}$ m.

## 8.2 The states of matter

The molecules in a solid are each anchored to one position, about which they vibrate continuously, as if held in position by a framework of springs. When heat energy is supplied, thus raising the temperature, the molecules vibrate faster and through greater distances than before. Thus the extra energy is transformed into kinetic energy of the molecules.

If sufficient heat energy is supplied a solid will melt to form a liquid. The amplitude of vibration of the molecules becomes so large that they break away from the position to which they were anchored and move freely amongst each other. However, forces still act between the molecules holding them close to each other and so although a liquid has no definite shape it does have a definite volume. The volume of a liquid is much the same as the solid from which it forms, thus the average separations of the molecules are about the same in each case, as are the densities. The molecular separation in the liquid may be slightly greater than in the solid (e.g. wax) or slightly less (e.g. water).

Not all the molecules have exactly the same energy at a particular temperature. Some molecules in a liquid have more than average energy and if they are near the liquid surface they are able to escape; this is evaporation. If heat energy is supplied to the liquid the average energy of each molecule increases; that is they move faster. Eventually all the molecules have sufficient energy to break away from each other and so a gas is formed, in which all molecules move independently. This change of state is known as boiling.

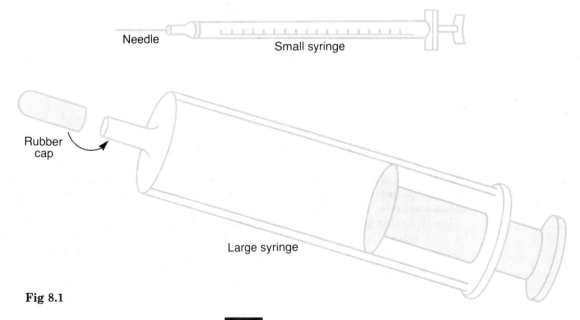

Fig 8.1

The molecules of a gas move continually, colliding with each other and with the walls of their containing vessel. The laws of mechanics apply to these collisions; further they are elastic, that is, no kinetic energy is lost. When a molecule bounces back from the walls of the container its momentum is changed and so it must have experienced a force. The force the molecules exert on the walls accounts for the pressure. If a gas is heated its molecules move faster and exert a greater pressure on the walls. An example of this effect is that if the pressure of a car or bicycle tyre is measured on a hot day it will be found to be greater than on a cold day, even though no gas has been allowed to enter or leave the tyre in the meantime. In a gas the forces between the molecules are so small that the molecules can be considered to move independently of each other. A gas therefore fills all the available space and has no fixed volume or shape; it takes the volume and shape of the vessel containing it. When the pressure exerted by a gas is equal to atmospheric pressure its volume will be about 1000 times greater than the liquid from which it forms, and its density therefore about 1000 times less. This can be shown using the syringes illustrated in Fig. 8.1.

A rubber cap is fitted over the end of the large syringe with its piston pushed down to zero volume. The small syringe with a hypodermic needle at its end is partially filled with water. 0.1 ml of water is then injected into the large syringe through the rubber cap via the hypodermic needle. When the needle is removed the cap seals. The large syringe is now inverted in a beaker of brine and the brine brought to the boil. The water in the syringe turns to steam and is seen to occupy a volume of about 100 ml; that is 1000 times larger than when it was water.

## 8.3 Brownian motion

The continual motion of molecules within a liquid or gas is called **Brownian motion**, after Robert Brown. In 1827 he used a microscope to examine pollen particles sprinkled on the surface of water and was surprised to notice that they were in a continuous state of haphazard movement. It appears that the motion of these relatively large particles is caused by the impact of moving water molecules. The same kind of movement can be seen in the case of smoke particles in air. The apparatus for this experiment is shown in Fig. 8.2.

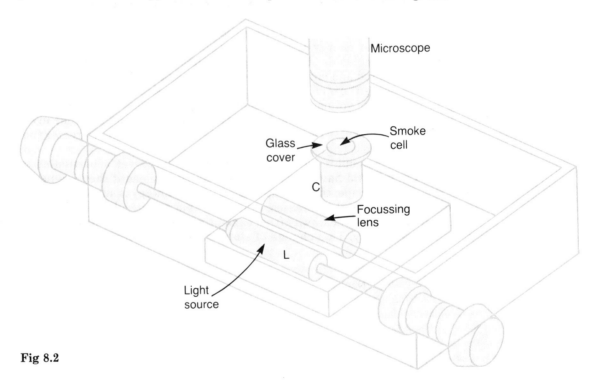

**Fig 8.2**

It consists of a small transparent cell $C$, with a cover, strongly illuminated from the side by a light $L$. A piece of cord or rag is set smouldering and some of the smoke which contains minute particles is collected by a syringe and injected into the cell. The cover is replaced and the microscope focused on the cell. The particles are seen to be moving in an irregular way, darting about suddenly, and always in motion.

The irregular motion of a particle is due to the movement of air molecules, which bombard it from all sides. The particle is relatively small and so the number of air molecules hitting it on one side is not balanced by an equal number hitting the opposite side at the same instant. The smoke particle thus moves in the direction of the resultant force. The irregular motion of the particles shows that air molecules move rapidly in all directions. At higher temperatures the molecules move even more rapidly and the motion of the particles is even more violent and irregular.

If the particles in a gas are very much bigger than the molecules of the gas they do not show this irregular movement. This is because the large number of molecules hitting one side is not relatively much greater than the number hitting the other at the same moment. The resultant force is thus relatively very small and the large particle is not so easily moved. For example, a table tennis ball suspended in air does not move for this reason.

## 8.4 Diffusion

Diffusion provides further evidence for the irregular random motion of molecules. Consider the apparatus shown in Fig. 8.3.

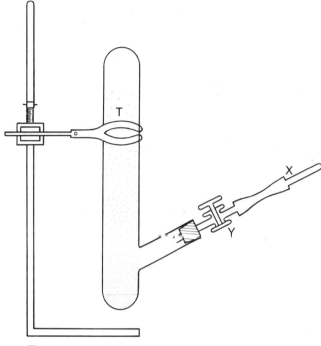

Fig 8.3

The vertical tube $T$ initially contains air. A capsule containing liquid bromine is placed in the end tube $X$ and this connected to the tap $Y$ by means of a short length of thick-walled rubber tubing. The tap is closed and the capsule tapped until it moves down inside the rubber tubing. The tubing is then squeezed with a pair of pliers until the capsule breaks. Liquid bromine runs down to the tap which is then opened allowing the bromine to enter the tube $T$. If the tube is observed some minutes later the characteristic brown colour of the bromine will be seen to have diffused part way up the tube. This is made more obvious by placing a white card behind the tube. If the experiment is now repeated, with the tube $T$ initially evacuated of air, the brown bromine vapour will fill the entire tube the moment the tap is opened.

The behaviour of the bromine in this experiment can only be explained by saying that the molecules of the bromine gas continually collide with air molecules in the first case. Their progress up the tube is thus impeded, whereas in the second case there are no air molecules to get in their way.

## 8.5 Randomness and mean free path

The continual collisions of molecules within a gas means that the path of an individual molecule is as shown in Fig. 8.4.

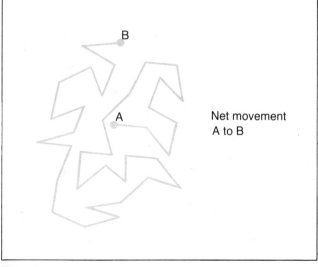

Fig 8.4

Although the actual speed of a molecule is very high (about 500 m/s or 1000 miles per hour) the net movement in one direction is very small because the molecule continually doubles back on its path. For this reason, if the stopper is removed from a bottle containing a pungent smelling liquid or gas, the smell will take some minutes to reach the far side of the room.

The average distance travelled by a molecule between collisions is termed **the mean free path**. Because of the **randomness** of collisions a molecule will travel different distances between collisions, but the average may be calculated (about $10^{-7}$ m at atmospheric pressure). The mean free path increases as the density of a gas is reduced.

## 8.6 Summary

1. Any material can exist in one of three states, as a solid, a liquid or a gas.
2. Energy is required to change a solid to a liquid and a liquid to a gas without a change in temperature.
3. The volume of any material is about the same when it is a liquid as when it is a solid.
4. When a material changes to a gas its volume increases about one thousand-fold.
5. The continual random motion of molecules within a liquid or gas is called **Brownian motion**.
6. Diffusion is the slow mixing of the molecules of one liquid or gas with another by the random motion of their molecules.

# 9 EXPANSION OF SOLIDS AND LIQUIDS

## 9.1 Solids

As temperature increases, the molecules of a solid vibrate with greater amplitude, thus causing the total volume of the solid to increase; that is, a solid expands when heated. This expansion can be troublesome and very large forces may be set up if there is an obstruction to the free movement of the expanding body. Continuous welded rails used on most of our railway lines are held in place by strong concrete clamps and sleepers. They are stress-free at 25°C but experience very large forces at high and low temperatures.

Modern motorways are often constructed with a concrete surface for economy. If this surface were laid in one continuous section cracks would appear owing to the expansion and contraction brought about by the differing summer and winter temperatures. To avoid this, the surface is laid in short sections, each section being separated from the next by a small gap filled with black pitch. On a hot day the pitch is squeezed out by the expansion of the concrete. Allowance has to be made for the expansion of bridges and the roofs of buildings made of steel girders. A common way of overcoming these difficulties is to fix one end of the structure while the other rests on steel rollers.

Although expansion can be troublesome it can be useful. Steel plates such as those used in ship building are often rivetted together using red hot rivets. Holes are made in the overlapping plates, a red hot rivet pushed through and its head pressed tightly against one plate. The other end of the rivet is hammered until it is tight against the second plate. As the rivet cools it contracts thus pulling the two plates together. A watertight join is formed.

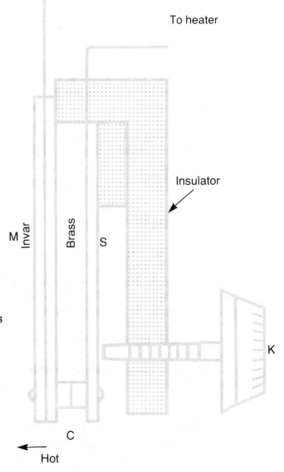

When strips of the same length but of different substances are heated through the same range of temperature, their expansions are not always equal. This difference may be used to make a thermostat. Figure 9.1 shows the principle of a thermostat which is a device for maintaining a steady temperature. The heater circuit is completed through the two contacts at C. One of the contacts is attached to the end of a metal strip S, the other to the end of a bimetallic strip M. On heating the brass expands more than the invar and the strip bends with the brass on the outside. At a certain temperature the strip bends so much that the contacts are pulled apart thus breaking the circuit to the heater. When the air cools the bimetallic strip straightens, the contacts close and the heater switches on again. The thermostat may be set to operate at different temperatures by adjusting the knob K. If the knob pushes the metal strip S towards the bimetallic strip a higher temperature is maintained. Thermostats working on the same principle are used to control the temperature of electric irons, immersion heaters, aquaria for fish and for other purposes.

Fig 9.1

## 9.2 Liquids

Liquids expand more than solids on heating. This may be shown by means of a flask fitted with a rubber bung and a length of glass tubing as shown in Fig. 9.2.

The flask is filled with a liquid and the bung pushed in until the level of liquid comes a short distance up the tube. When the flask is plunged into hot water the liquid level is first seen to fall and then to rise.

The initial fall is due to the expansion of the glass which becomes heated first and expands before the heat has had time to reach the liquid.

Due to the expansion of the glass, the expansion of the liquid measured is less than its true expansion.

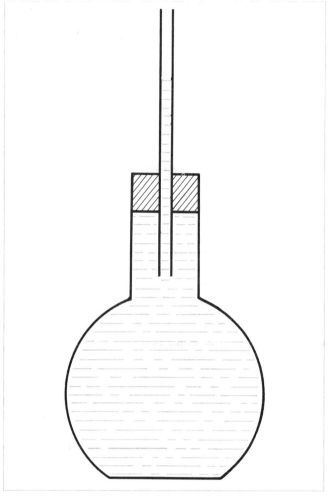

Fig 9.2

If we take some water at 0°C and begin to heat it the water contracts instead of expanding over the temperature range 0°–4°C. Its behaviour is unusual. At about 4°C the water reaches its smallest volume, and thus its greatest density. Above 4°C water expands in the normal way. The unusual behaviour of water means that between 0° and 4°C water is less dense the colder it is. The colder water rises to the surface of a pond and it is here that ice first forms. A sheet of surface ice acts as a heat insulator and it takes a long time for the water below to freeze, thus thickening the ice. It is most unlikely that a pond of reasonable depth will freeze right through; thus pond life is able to survive below the ice.

## 9.3 Thermometers

We must be careful to distinguish the temperature of a body from the internal energy it contains. Temperature is a measure of the degree of hotness of a body whereas the energy it contains depends on its nature and mass as well as its temperature.

The temperature of a substance is a number which expresses its degree of hotness on some chosen scale. It is measured by means of a thermometer. Some thermometers depend on the expansion of a liquid when heated, some on the expansion of a bimetallic strip, and others on the change of other physical quantities brought about by heating; for example electrical resistance.

The most common thermometer in use is that which relies on the expansion of mercury in a glass tube when the mercury is heated. The mercury is contained in a bulb at the lower end of the tube. Above the mercury is some nitrogen. The glass tube is calibrated by dividing it into 100 equal divisions between two fixed points. The upper fixed point is marked on the tube when the thermometer is surrounded by steam boiling under standard atmospheric pressure. The lower fixed point is marked when the thermometer is in pure melting ice. The upper and lower fixed points are given the numbers 100 and zero, respectively. This procedure establishes the Centigrade or Celsius scale of temperature.

Mercury freezes at $-39°C$ and therefore a mercury in glass thermometer is not suitable for use in countries such as Russia and Canada which have very cold winters. In such places alcohol in glass thermometers are used, as alcohol remains liquid down to $-115°C$. However, alcohol boils at a little above 70°C and so is not suitable for high temperatures.

A clinical thermometer is one specially designed to measure the temperature of the human body. It is only necessary for it to have a range of a few degrees on either side of normal body temperature (37°C). The thermometer is generally placed beneath the patient's tongue and left there for two minutes to ensure it acquires the body temperature. The stem has a narrow constriction in its bore just above the bulb (Fig. 9.3). Thus when the thermometer is removed

Fig 9.3

from the mouth, the mercury beyond the constriction stays put while that below it contracts into the bulb. When the temperature reading has been taken, the mercury in the tube is returned to the bulb by shaking.

## 9.4 Summary

1. Solids and liquids both expand on heating (water only above 4°C).
2. A thermostat relies on the different amounts of expansion of the two metals in the bimetallic strip for its operation.
3. Thermometers make use of the expansion of mercury or alcohol which occurs on heating.

# 10 THE BEHAVIOUR OF GASES

The volume of a gas can be changed not only by altering its temperature, but also by changing the pressure exerted on it. Thus a gas has three quantities: **volume**, **temperature** and **pressure** all of which may change. In order to make a full study of the behaviour of a fixed mass of gas three separate experiments are therefore carried out to investigate:

1 the relation between volume and pressure at constant temperature (**Boyle's law**);
2 the relation between volume and temperature at constant pressure (**Charles' law**);
3 the relation between pressure and temperature at constant volume (**pressure law**).

## 10.1 Boyle's law

**The volume of a fixed mass of gas is inversely proportional to the pressure, provided the temperature remains constant; that is, the pressure multiplied by the volume is constant.**

$$pV = \text{constant}$$

One version of the apparatus used to show this law is illustrated in Fig. 10.1. It consists of a column of air trapped in a vertical tube by some oil with a low vapour pressure. Pressure is applied to the oil in the reservoir by a pump. The Bourdon gauge measures the pressure of the air above the oil in the reservoir. This is a little greater than the pressure of the air trapped in the tube, due to the vertical oil column, but the error is so small that for practical purposes it may be ignored.

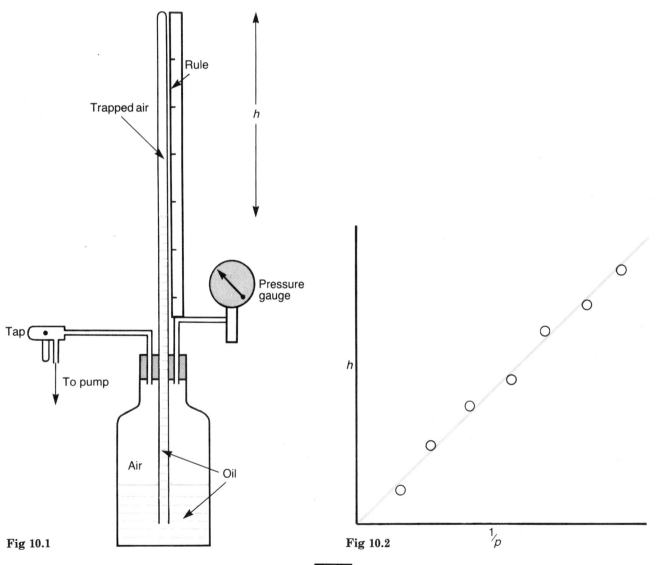

Fig 10.1

Fig 10.2

Air is first pumped into the reservoir until the Bourdon gauge reaches its maximum reading. The tap is closed and readings taken of the length $h$ of the trapped air column and also the pressure reading of the Bourdon gauge $p$. The tap is then opened to allow a little air to escape, closed again, and a further set of readings recorded. This procedure is repeated until the Bourdon gauge registers atmospheric pressure once more. It is possible, using a suction pump, to obtain readings below atmospheric pressure. If $h$ is now plotted against $1/p$ a graph is obtained similar to that in Fig. 10.2.

As the volume $V$ is proportional to the length of the column of trapped air $h$, the fact that the graph is a straight line through the origin shows that:

$$V \div 1/p = \text{a constant}$$
$$\text{or} \quad pV = \text{a constant}$$
$$\text{or} \quad \boldsymbol{p_1 V_1 = p_2 V_2} \text{ etc.} \tag{10.1}$$

## 10.2 Charles' law

**The volume of a fixed mass of gas is directly proportional to its absolute temperature provided the pressure remains constant; that is, the volume divided by the absolute temperature is constant.**

$$\frac{V}{T} = \text{constant}$$

A simple apparatus for verifying this law is shown in Fig. 10.3. It consists of a capillary tube sealed at its lower end. Some air has been trapped in the tube by a short thread of mercury $M$. A centimetre scale is attached to the tube so that the length of the trapped air column can be easily noted. The capillary tube and scale are placed in a beaker of water alongside a thermometer $T$.

The temperature of the water and the length of the air column are first noted. The water is then heated through about 10°C, time allowed for the heat to reach the air, and the temperature and length of the air column recorded. This process is repeated several times. The volume $V$ of trapped air is proportional to the length of the column as the tube is of uniform

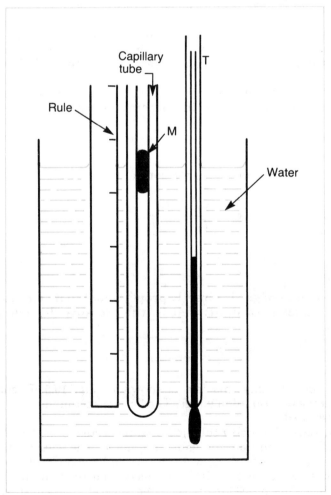

**Fig 10.3**

bore. The trapped air is kept at constant pressure by the mercury index moving up the tube as the temperature increases.

From the results obtained a graph of length against temperature is plotted (Fig. 10.4). The graph is a straight line showing that air expands uniformly with temperature as measured on the mercury thermometer.

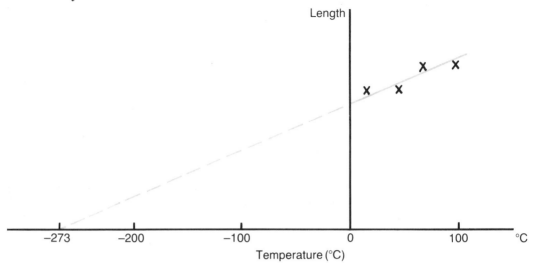

**Fig 10.4** Variation of volume with temperature at constant pressure

If the graph is produced backwards it cuts the temperature axis at a point which gives the temperature ($-273°C$) at which the volume of the gas would contract to zero, assuming the gas continues to contract uniformly below $0°C$. As we cannot imagine it possible for a gas to have a volume of less than zero, it is reasonable to assume that $-273°C$ is the lowest temperature it is possible to obtain, and thus represents the absolute zero of temperature. This assumption cannot be directly tested by experiment as gases liquefy before they reach this temperature, and so the gas laws no longer apply. However, experiments have shown that while temperatures close to $-273°C$ have been reached, it has not been possible to go below this value.

The value $-273°C$ has thus been taken as the zero of a new scale of temperature called the **Absolute** or **Kelvin** scale. Temperatures on this scale are represented by $T$, and are expressed in units of K. Temperatures on the Celsius scale are converted to the Kelvin scale by adding 273. Thus $0°C = 273$ K.

The graph in Fig. 10.4 is a straight line through the origin of our new temperature scale. Thus the volume of the gas is proportional to its temperature measured on the Kelvin scale and we may write:

$$V \propto T$$

or $\quad \dfrac{V}{T} = $ a constant $\hfill (10.2)$

or $\quad \dfrac{V_1}{T_1} = \dfrac{V_2}{T_2}$ etc.

## 10.3 Pressure law

**The pressure of a fixed mass of gas is directly proportional to its absolute temperature provided its volume remains constant; that is, the pressure divided by the absolute temperature is constant.**

$$\frac{p}{T} = \text{constant}$$

The apparatus for demonstrating this law is shown in Fig. 10.5. It consists of a flask connected by rubber tubing to a Bourdon gauge. The flask is surrounded by water in a beaker. A thermometer $T$ is also used.

The apparatus, particularly the rubber tubing, is first inspected for leaks. The temperature of the water and the reading on the Bourdon gauge are noted. The water is then heated while being stirred. After the temperature of the water has risen by about $10°C$, heating is stopped, time allowed for the heat to reach the air in the flask, and then the temperature and the pressure are again noted. This procedure is repeated several times until the water is near boiling.

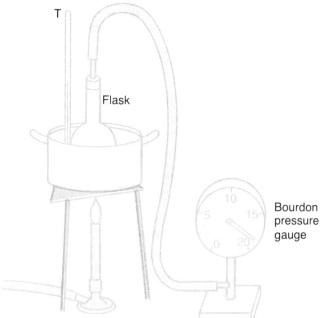

**Fig 10.5**

From the results a graph of pressure against temperature is plotted (Fig. 10.6). It is a straight line passing through $-273°C$ or $0$ K. Thus the pressure of the gas is proportional to its temperature measured on the Kelvin scale and we may write:

$$p \propto T$$

or $\quad \dfrac{p}{T} = $ a constant $\hfill (10.3)$

or $\quad \dfrac{p_1}{T_1} = \dfrac{p_2}{T_2}$ etc.

This apparatus can be used to measure temperature (constant volume gas thermometer).

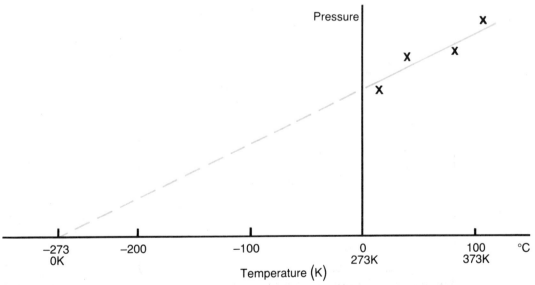

**Fig 10.6** Variation of pressure with temperature at constant volume

## 10.4 The universal gas law

The equations (10.1), (10.2) and (10.3) can be combined in a more general equation which we may write:

$$\dfrac{pV}{T} = \text{a constant} \hfill (10.4)$$

or $\quad \dfrac{p_1 V_1}{T_1} = \dfrac{p_2 V_2}{T_2}$ etc. $\hfill (10.5)$

If we always consider one mole (one gram molecular weight) of a gas, the constant is the same for all gases and we may write equation (10.4) as follows:

$$\frac{pV}{T} = R$$

where $R$ is the universal gas constant.

Suppose a fixed mass of gas, occupying 1 litre at 27°C, is heated to 227°C and at the same time the pressure on the gas is doubled. We may find its final volume by using equation (10.5), but first we must convert the temperatures to the Kelvin scale.

$$27°C = 300 \text{ K} \quad \text{and} \quad 227°C = 500 \text{ K} \quad \text{and} \quad p_2 = 2p_1$$

$$\frac{p_1 V_1}{T_1} = \frac{p_2 V_2}{T_2}$$

thus $\quad \dfrac{p_1 \times 1}{300} = \dfrac{2p_1 \times V_2}{500}$

hence $\quad V_2 = \dfrac{500}{300 \times 2} = \dfrac{5}{6}$ litre

## 10.5 Models of a gas

The following models give an idea of how we think molecules of a gas behave. The first model consists of several marbles placed on the base of a tray or baking tin. The tray is moved horizontally by hand in short, sharp random jerks. The marbles are seen to move, making random collisions with each other and the sides of the tray. To obtain a clear impression it is best to watch one marble carefully.

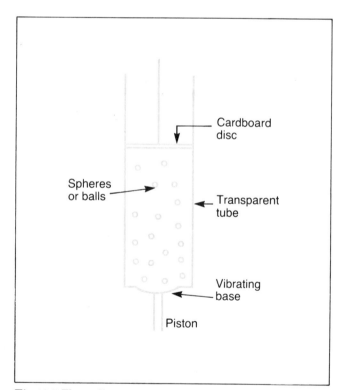

**Fig 10.7** Three-dimensional model of a gas

Although this model is useful, it is two-dimensional, whereas molecules move in three dimensions. Figure 10.7 shows a better model. Some small polystyrene spheres or phosphor bronze ball bearings are contained in a vertical transparent tube. The base of the tube is connected to a piston which is driven by an electric motor. The oscillating piston causes the base to vibrate; the frequency of vibration is changed by altering the speed of the motor. An alternative arrangement is to use a loudspeaker as the base of the tube and to connect the loudspeaker to a signal generator the frequency of which can be altered.

When the base vibrates, the spheres on it are thrown into random motion in the cylinder. They collide with each other and with the walls of the container on which they thus exert a pressure. They also exert a pressure on the cardboard disc which sits on top of them. If the amplitude of vibration of the base is increased, the average energy of the spheres is increased.

The spheres thus move faster and keep the disc at a higher level. This illustrates the expansion of a gas when it is heated at constant pressure. Instead of allowing the disc to rise when the spheres are made to move faster, masses can be added to the top of the disc to keep it at the same height. The masses increase the external pressure that the disc exerts on the spheres. This illustrates the increase in the pressure that a gas exerts when it is heated at constant volume.

## 10.6 Summary

1 For a fixed mass of gas the following equation is true:

$$\frac{pV}{T} = \text{a constant}$$

where $p$ is the pressure of the gas, $V$ the volume that it occupies and $T$ its temperature on the Kelvin scale (°C + 273).

2 If one mole of gas is used the equation becomes

$$\frac{pV}{T} = R$$

where $R$ is the universal gas constant.

3 If one of the quantities $p$, $V$ or $T$ is kept constant three simpler equations may be used. They are

$$pV = \text{a constant, when } T \text{ is kept constant}$$

$$\frac{p}{T} = \text{a constant, when } V \text{ is kept constant}$$

$$\frac{V}{T} = \text{a constant when } p \text{ is kept constant}$$

# 11 SPECIFIC HEAT CAPACITY

Heat is a form of energy and like any other form of energy it is measured in joules. The size of one joule is defined in terms of mechanical units.

If we take equal masses of water and oil and warm them for the same time in separate containers using similar immersion heaters, the temperature of the water will rise much less than the temperature of the oil. We say these substances have different specific heat capacities.

**The specific heat capacity of a substance is the quantity of heat required to raise the temperature of one kilogram of it by 1 K.** It has units of joules per kilogram per Kelvin (J/kg K).

It follows that if $m$ kilograms of a substance, of specific heat capacity $c$, are to be raised in temperature by $\theta$ K, then the heat required will be $mc\theta$ joules.

$$\text{Heat required} = mc\theta \text{ joules}$$

It happens that for water 4200 joules of heat are required to raise the temperature of one kilogram by 1 K. The specific heat capacity of water is thus 4200 J/kg K. This value is high compared with most other substances. A great deal of energy is required to raise the temperature of water a certain amount compared with the same mass of another substance. Likewise water cools more slowly because it contains more energy than the same mass of other substances at the same temperature. Water is therefore used to fill radiators and hot water bottles.

## 11.1 To measure the specific heat capacity of a solid

The temperature of an object can be raised by supplying energy to it in various ways.
1. Mechanical energy can be supplied by allowing the object to fall through a height $h$, when the energy supplied is equal to $mgh$.
2. The energy can be obtained from another hot body of mass $m$, and specific heat capacity $c$. The energy supplied is equal to $mc\theta$, where $\theta$ is its temperature change.
3. The energy can be supplied by an electric current $I$, flowing for a time $t$, through a wire across which a potential difference of $V$ is maintained (see Unit 21.7).

$$\text{Energy supplied} = VIt$$

The mechanical method may be used to determine the specific heat capacity of lead shot. The temperature of about 0.25 kg of lead shot is recorded. It is then placed in a wide tube about one metre long, and rubber bungs fitted at each end of the tube. The tube is smartly inverted about 100 times, the number $n$ of inversions being counted. One bung is removed from the tube and the temperature of the lead shot again measured. The specific heat capacity $c$ is calculated from the equation:

$$n \times mgh = mc\theta$$

If two substances at different temperatures are mixed and come to a common temperature, the heat lost by one in cooling will be equal to the heat gained by the other, providing no heat is gained from or lost to the surroundings.

The specific heat capacity of a solid can be found by warming it to a high temperature and then quickly transferring it to a calorimeter containing cold water. The water and calorimeter receive heat from the solid and all three finally reach the same temperature. The calorimeter should either be made from a material which is a very poor conductor, in which case it can be assumed to take no heat, or from a material such as copper which is such a good conductor that it can be assumed to have the same temperature as its contents.

Details of an experiment to find the specific heat capacity of copper using a copper calorimeter are as follows. A large piece of copper, with a thread attached to it, is placed in a beaker of boiling water and left for some time. While the copper is warming to 100°C, a copper calorimeter is weighed empty and then about two-thirds full of cold water. The calorimeter is then placed inside a jacket and a thermometer placed in it.

When the piece of copper has been in the boiling water long enough to reach 100°C, the temperature of the cold water is noted, and the copper transferred from the boiling to the cold water. The mixture is gently stirred by moving the piece of copper in the water by means of the thread. The final steady maximum temperature is noted. The piece of copper is then dried and weighed.

The heat lost by the solid is then equated to the heat gained by the calorimeter and the cold water. The specific heat capacity of copper is the only unknown in the equation and can be calculated.

*Example*

Suppose the following readings have been obtained:

| | |
|---|---|
| mass of calorimeter empty | $m_1 = 0.1$ kg |
| mass of water in calorimeter | $m_2 = 0.1$ kg |
| mass of piece of copper | $m_3 = 0.2$ kg |
| temperature of cold water | $\theta_1 = 10°C$ |
| temperature of boiling water | $\theta_2 = 100°C$ |
| temperature of mixture | $\theta_3 = 20°C$ |
| specific heat capacity of water | $c_1 = 4200$ J/kg K |

Heat gained by calorimeter and water = heat lost by piece of copper. Thus

$$m_1 c(\theta_3 - \theta_1) + m_2 c_1(\theta_3 - \theta_1) = m_3 c(\theta_2 - \theta_3)$$

hence $\quad 0.1c \times 10 + 0.1 \times 4200 \times 10 = 0.2c \times 80$

and $\quad\quad\quad\quad\quad\quad\quad\quad 15c = 4200$

$$c = 280 \text{ J/kg K}$$

Although the method just described is a reasonable one, it is open to two serious errors, some hot water is carried across to the calorimeter on the hot solid and the hot solid loses heat during the transfer. Both errors can be reduced with care but not eliminated.

A better method is one based on the steady heating of a block of the solid by a heater immersed in it, as shown in Fig. 11.1.

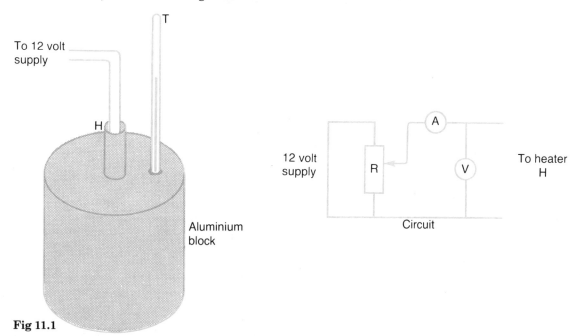

**Fig 11.1**

A 12 V immersion heater $H$, with a power of 24 or 36 W, is sunk into a hole specially drilled in the solid for this purpose. It is convenient, but not essential, if the block has a mass of 1 kg. A second, smaller hole in the block contains a thermometer $T$. To ensure good thermal contact between the heater and block and the thermometer and block, a few drops of thin oil are placed in each hole. The block can be surrounded by an insulating jacket to reduce heat losses.

The apparatus is connected up and switched on for a known time (between 10 and 30 minutes depending on the material of the block). During this time the voltage and current are kept as constant as possible, by adjusting the variable resistor $R$, and their values noted. At the beginning and end of the experiment the temperature of the block is noted. The specific heat capacity $c$ of the material of the block is worked out using the equation:

energy supplied electrically = heat gained by the block,

$$VIt = mc(\theta_2 - \theta_1)$$

where $\theta_1$ and $\theta_2$ are the initial and final temperatures of the block.

As well as avoiding the errors present in the method of mixtures, this method requires no calorimeter.

*Specific heat capacity*

## 11.2 To measure the specific heat capacity of a liquid

If the specific heat capacity of a solid is known, the method of mixtures may be used to find the specific heat capacity of a liquid. The calorimeter is two-thirds filled with the liquid under test instead of water. The procedure and calculations are then the same as already described for finding the specific heat capacity of a solid.

However, the electrical heating method just described for a solid is again superior. The liquid (1 kg for convenience) is contained in an aluminium saucepan. The heater is immersed in the liquid, care being taken not to short the connections. The procedure is the same as that described for a solid. Strictly speaking, in working out the specific heat capacity of the liquid, account should be taken of the heat used in raising the temperature of the aluminium saucepan. However, both the mass and specific heat capacity of aluminium are likely to be much less than the values for the liquid, so no great error is involved in ignoring the heat taken in by the saucepan. Calculation of the result is then the same as for a solid.

## 11.3 Summary

1 The specific heat capacity of a substance is the quantity of heat required to raise the temperature of one kilogram of it by 1 K. It has units of joules per kilogram per degree Kelvin (J/(kg K)).

2 The heat required to change the mass $m$ of a substance by a temperature of $\theta$ degrees is given by

$$\text{heat required} = mc\theta \text{ joules}$$

where $c$ is the specific heat capacity of the substance.

3 The specific heat capacity of a solid or liquid is usually measured electrically, the electrical energy supplied being equal to $VIt$ joules.

# 12 LATENT HEAT

## 12.1 Vaporization

When water is heated at atmospheric pressure it begins to boil at 100°C. Once boiling begins the temperature of the water remains constant at 100°C, even though heating continues. The water is steadily absorbing energy from the heater, yet there is no increase in temperature. This energy is the energy needed to convert the water from the liquid state to the vapour state. It is used in freeing the molecules from the influence of other molecules, so that they now move independently. It is given the name **latent heat**; latent means hidden or concealed.

**The specific latent heat of vaporization of a substance is the quantity of heat required to change unit mass of the substance from the liquid to the vapour state without change in temperature.**

Its value for water at 100°C is $2.26 \times 10^6$ J/kg.

## 12.2 Fusion

Just as latent heat is taken in when water changes to steam at the same temperature, so the same thing happens when ice melts to form water.

**The specific latent heat of fusion of a substance is the quantity of heat required to convert unit mass of the substance from the solid to the liquid state without change in temperature.**

Its value for ice at 0°C is $3.34 \times 10^5$ J/kg.

It follows from both definitions that if $m$ kilograms of a substance, of specific latent heat $L$, change from one state to another without change in temperature, then the heat absorbed or given out will be $mL$ joules. Thus

$$\text{energy change} = mL \text{ joules}$$

## 12.3 To measure the specific latent heat of fusion of ice

A 12 V immersion heater should be placed in a filter funnel and surrounded by closely packed dry ice (Fig. 12.1).

A voltmeter and ammeter are connected in the supply circuit so that the energy supplied can be determined.

The immersion heater is switched on for a known time (say 3 minutes) and the mass of water produced is found by weighing the beaker before and after the experiment.

A second 'control' apparatus is set up alongside and water collected from this during the same time.

This tells us the mass of ice melted in the main experiment by absorption of heat from the surroundings.

The difference between the two masses collected gives the true mass melted by the heater. $L$ is calculated using the equation:

$mL = VIt$

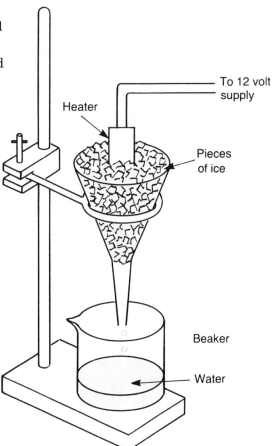

**Fig 12.1** Specific latent heat of fusion of ice

## 12.4 To measure the specific latent heat of vaporization of water

A tall beaker of water is placed on a top pan balance. A 240 V mains-operated immersion heater is carefully suspended in the water so that it does not touch the bottom of the beaker, yet is well covered with water. The heater is switched on and when the water is boiling briskly a note is made of the time and the reading of the balance. The heater is left running until a few hundred grams of water have vaporized; the time this takes will depend on the power of the heater. When sufficient water has vaporized the time and balance reading are again noted. As the immersion heater is mains operated, the voltage will be steady and dependable and the power provided by the heater will be that stated on it. The specific latent heat $L$ is calculated from the equation:

$$mL = VIt$$

where $m$ is the mass of water vaporized.

## 12.5 Evaporation

**Evaporation** takes place from the surface of a liquid at all temperatures, whereas boiling only occurs above a certain temperature and takes place throughout the liquid. However, the nearer the liquid is to its boiling temperature the faster the rate of evaporation.

Evaporation means that the faster molecules which happen to be near the surface escape from it. If the molecules which escape are free to move away from the space immediately above the liquid (or even encouraged to do so by a stream of air) evaporation will continue until the liquid has all evaporated. This is how puddles of water left on the road after rain eventually dry up. The higher the temperature of the liquid the quicker evaporation occurs.

As evaporation means that some of the more energetic molecules of the liquid are leaving it, the average energy of the molecules left behind falls; thus the liquid falls in temperature. For example, a bottle of milk may be kept cool by wrapping the bottle in a wet cloth. Some water evaporates from the cloth and the remaining water falls in temperature. In turn it extracts heat from the milk. This is more effective if the rate of evaporation can be speeded up by standing the wet bottle in a draught.

It is unwise for a human being to stand in a draught or breeze after taking violent exercise, however warm he feels. The perspiration on his body evaporates quickly under these conditions, thus cooling his body, and making it susceptible to a chill.

## 12.6 Summary

1. Energy is required to change a solid to a liquid or a liquid to a gas even though no change in temperature occurs. This energy is given the name latent heat.
2. The latent heat of fusion of a substance is the quantity of heat required to change one kilogram of it from solid to liquid without a change in temperature.
3. The latent heat of vaporization of a substance is defined in a similar way for a change from liquid to gas.

# 13 TRANSMISSION OF HEAT

Heat is transferred from one place to another in one of three ways: **conduction**, **radiation** or **convection**.

## 13.1 Conduction

If a metal spoon is left in a teacup for a short length of time the handle becomes warm. Heat travels along the spoon by means of conduction. Metals contain electrons which are very loosely attached to atoms, and are easily removed from them. When a metal is heated these 'free electrons' gain kinetic energy and move independently of the atoms. They drift towards the cooler parts of the metal thus spreading the energy to those regions.

In substances where no free electrons are present the energy is conveyed from one atom to another by collision. This is a much slower process and such substances are called poor conductors of heat.

Most metals are good conductors of both heat and electricity: free electrons being responsible for both. Substances such as wool, cotton, cork and wood are bad conductors. A number of materials lie between these extremes. The best saucepans are made of copper as heat is rapidly conducted through this metal. Many materials are poor conductors of heat because they trap tiny pockets of air between their fibres, and air, like all gases, is a poor conductor of heat. Textiles and glass fibre are examples. Glass fibre is frequently used as lagging for attics and hot water tanks.

## 13.2 Comparison of thermal conductivities

Similar rods, made of different materials, pass through corks inserted in holes made in the side of a metal tank (Fig. 13.1). The rods are first coated in wax by dipping them into molten wax and then allowing it to cool. Boiling water is then poured into the tank, so that the ends of the rods are all heated to the same temperature. After some time it is seen that the wax has melted to different distances along the rods, showing differences in their **thermal conductivities**.

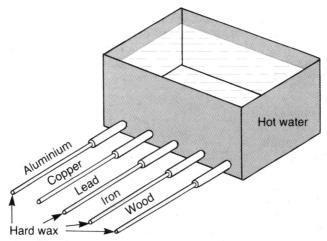

**Fig 13.1** Conduction

In the kitchen, saucepans are made of metals such as copper or aluminium, which are good heat conductors. However, the handles of saucepans are made of insulators, such as plastic or wood, so that the utensils can be handled when hot. The handles of kettles and oven doors must be made of similar materials.

## 13.3 Radiation

Both conduction and convection are ways of conveying heat from one place to another which require the presence of a material. Radiation does not require a material medium; it is the means by which heat travels from the sun through the empty space beyond the earth's atmosphere. Radiation consists of electromagnetic waves which pass through a vacuum. On striking a body these waves are partly reflected and partly absorbed, and can cause a rise in temperature.

The rate at which a body radiates depends on its temperature and the nature of its surface. For a given temperature, a body radiates most energy when its surface is dull black and least when its surface is highly polished. A comparison of the radiating powers of different surfaces may be made using a Leslie cube. This is a hollow metal cube, each side of which has a different surface; one is dull black, one highly polished, another may be shiny black and the fourth painted white. The cube is filled with hot water and a thermometer with a blackened bulb placed at the same distance from each face in turn (Fig. 13.2). In each case the thermometer reading is noted.

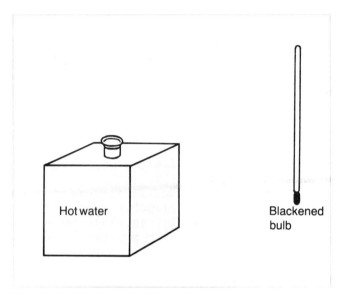

**Fig 13.2** Radiation — Leslie's cube

The results show that the dull black surface produces the highest reading and the highly polished one the lowest. This indicates that the dull black surface is the best radiator and the highly polished one the worst.

The absorbing powers of different surfaces may be compared using a small electric fire (Fig. 13.3). The fire is placed about 10 to 15 cm behind a heat insulating screen, faced with a polished metal surface towards the fire. The screen has a hole in it 2 or 3 cm in diameter. If a piece of aluminium foil is attached to the back of one's hand by damping it, it is found that the hand may be held over the hole in the screen without discomfort. If the foil is now painted over with lamp black (or matt black paint) and the hand again placed over the hole, the hand has to be removed after a few seconds or it becomes burnt. This shows that a dull black surface is a good absorber of heat as well as a good radiator, whereas a highly polished surface is poor in both respects.

Radiation from the sun is mostly in the form of visible light and infrared rays. These pass through glass and hence may reach the ground and plants inside a greenhouse, which absorb them. These objects also radiate, but due to their relatively low temperature, the infrared rays they emit are of longer wavelength and cannot penetrate the glass. The energy is thus trapped inside the greenhouse.

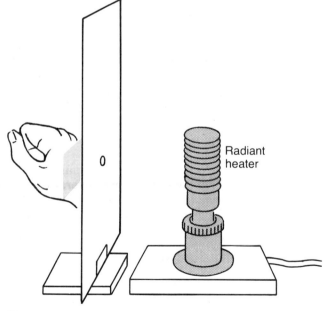

**Fig 13.3** Radiation

## 13.4 Convection

When a vessel containing a liquid is heated at the bottom, the liquid in that region becomes warm, less dense, and as a result rises. Its place is taken by cooler, more dense, liquid moving downwards. In **convection** the heat is carried from one place to another by the movement of the molecules of the liquid. The existence of convection currents may be shown by dropping a large crystal of potassium permanganate to the bottom of a beaker of water. The beaker is gently heated under the crystal, which should be near the centre of the base. An upward current of coloured water will rise from the place where the heat is applied. This current spreads out at the surface and then moves down the sides of the beaker (Fig. 13.4). A domestic hot water system relies on convection currents for its functioning.

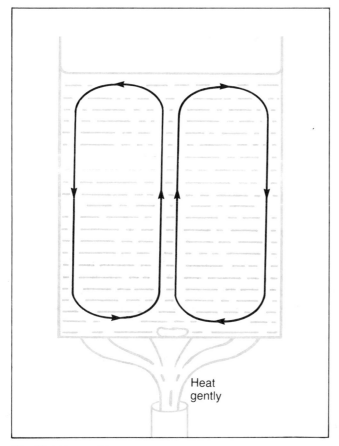

**Fig 13.4** Convection

Convection currents exist in gases as well as liquids. Warm air, for example, is less dense than cold air and rises. Cooler air then moves in to replace it. It is this process which is responsible for sea breezes towards the end of a warm summer's day. The relatively dark land absorbs more heat than the sea. Its specific heat capacity is less than that of water. As a result of both of these facts it reaches a higher temperature than the sea towards the end of a summer's day. The land warms the air over it which rises and is replaced by the cooler air from over the sea. Late on a clear summer's night the reverse is likely.

## 13.5 Summary

1. Heat is transferred from one place to another by conduction, convection or radiation.
2. Conduction takes place in solids. In good conductors such as metals 'free electrons' gain kinetic energy and move independently of the atoms. They move to the cooler parts of the solid thus spreading the energy to those regions. In substances where no 'free electrons' are present the energy moves from one atom to another by collision. This is a much slower process and such substances are called poor conductors of heat.
3. Radiation takes place by the transmission of waves from a hot body to any other object. Radiation can take place through a vacuum as in the case of heat from the sun reaching the earth.
4. When liquids or gases are heated their density decreases and the warm liquid or gas rises. Cooler liquid or gas falls to take its place. Convection currents are set up and heat is transmitted from one place to another.

# 14 INTRODUCTION TO WAVES

## 14.1 Progressive waves

The idea of **progressive waves** is best illustrated using a spiral spring or slinky stretched on a bench (Fig. 14.1). If one end of the spring is shaken at right angles to its length (Fig. 14.1(a)) a wave is seen to travel along the spring. As the wave is moving it is said to be progressive. In this example the motion causing the wave is at right angles to the direction of travel of the wave and the wave is termed **transverse**. In Fig. 14.1(b) one end of the spring is shaken in the direction of the spring's length and a concertina effect travels the length of the spring. In this case the motion causing the wave is in the same direction as the wave travels and the wave is called **longitudinal**.

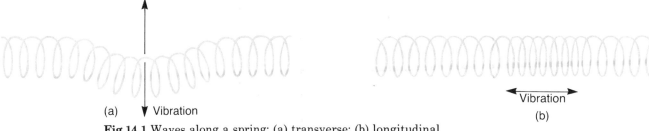

**Fig 14.1** Waves along a spring: (a) transverse; (b) longitudinal

Examples of transverse waves are television, radio, heat, light, ultraviolet, X-rays, $\gamma$-rays (all electromagnetic waves) and water waves. In electromagnetic waves the electric and magnetic disturbances are at right angles to the direction the wave travels. At any moment a transverse wave has the shape of a sine wave and can be drawn as such (Fig. 14.2).

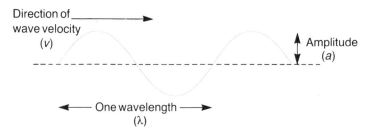

**Fig 14.2** Sine curve representing wave motion

Sound, on the other hand, is a longitudinal wave motion. The molecules of the material through which the sound travels, vibrate to and fro along the direction of motion of the wave. At a given time each molecule is at a different point in its motion. A longitudinal wave may be represented by a sine curve, but it must be remembered that although the $y$ direction still represents the size of the displacement, the direction of the displacement is in fact parallel to the direction of travel of the wave.

In Fig. 14.2 the **amplitude of the wave** is denoted by **a**. The amplitude represents the maximum displacement of the wave from the zero position. The distance between corresponding points on two successive waves is known as the **wavelength $\lambda$**. The number of waves produced every second by the source is called the **frequency $f$**. The length of wave motion produced every second by the source is the number of complete waves produced ($f$) multiplied by the length of each wave ($\lambda$). This product is clearly the speed $v$ of the wave motion. Thus

$$v = f\lambda$$

It is usual to measure the speed in metres per second, the frequency in hertz and the wavelength in metres.

## 14.2 The ripple tank

A good understanding of the properties of all types of waves can be obtained from studying the behaviour of water waves in a ripple tank. The construction of a ripple tank is shown in Fig. 14.3.

*The ripple tank* 51

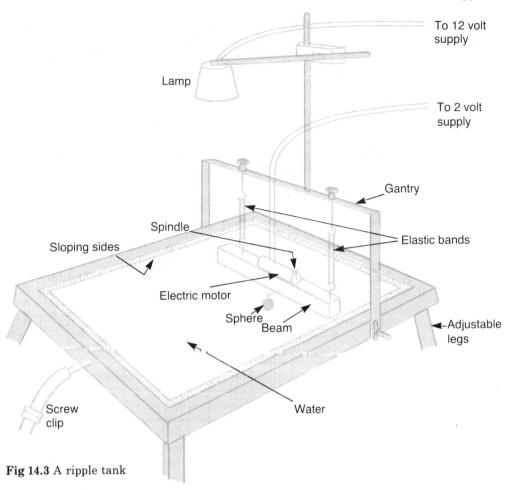

**Fig 14.3** A ripple tank

The apparatus consists of a shallow tray with sloping sides to reduce reflection. The tray is mounted on four legs, each of which can be adjusted in length by a screw foot. In this way the tray can be levelled. A gantry stands above one end of the tank. On this gantry is a post on which a lamp is fixed. A beam, with a motor mounted on its top, is suspended from the gantry. A number of spheres, each of which may be lowered to project below the beam, are fixed to it.

The behaviour of the waves is best seen by using the lamp to cast a shadow of the water surface on a sheet of paper at the feet of the tank. Wave peaks and troughs are clearly seen on the paper. In diagrams lines are drawn to represent the peaks or troughs. These lines may be considered to represent the surface over which energy from the source spreads out, and are called **wavefronts**.

Two simple types of wave formation are worth studying; straight and circular waves. A small number of straight waves may conveniently be made by rolling a short length of dowel rod to and fro on the bottom of the tank. A few circular waves may be produced by touching the water surface with a pencil the required number of times.

For many experiments it is necessary to produce a continuous series of waves of one or other type.
This is done by connecting the motor to a suitable power unit (often 2 V d.c.) whereupon the spindle which is unevenly weighted, sets the beam bouncing up and down.

The speed of the motor determines the frequency of the waves.
If straight waves are required the lower edge of the beam should be a few millimetres below the water surface; if circular waves are wanted then the beam should be raised clear of the surface but one or more spheres turned down so that they are partly submerged by the same amount.

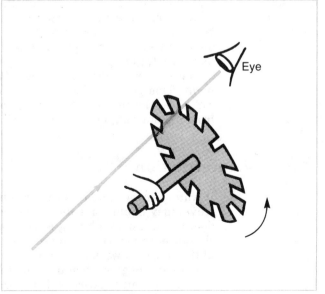

**Fig 14.4** Hand stroboscope

When the motor is producing a continuous series of waves it is often very difficult, if not impossible, for the eye to follow them across the paper below the tank. If this is so, it is helpful to use a hand stroboscope to 'freeze' the picture of the waves. A typical stroboscope is shown in Fig. 14.4. The handle is held in one hand while the disc is rotated using one finger of the other hand. The rotating disc is placed between the face and the wave pattern on the paper. The speed of rotation of the disc is increased until the waves appear stationary.

The slits in the disc allow glimpses of the waves at equal intervals of time. At one particular speed of rotation each wave moves forward, to the position of the wave in front of it, in between successive glimpses. The wave pattern thus appears stationary. If the stroboscope is rotated slightly more slowly the wave pattern appears to be moving slowly forward. The opposite is the case if the disc is rotated slightly too fast.

## 14.3 Reflection

To see this effect clearly only a small number of waves are required. Therefore it is best if the ripple tank motor is not used; the waves being produced in the way already described. Figure 14.5 shows the results of the incidence of both straight and circular waves on straight and circular reflecting barriers $R$.

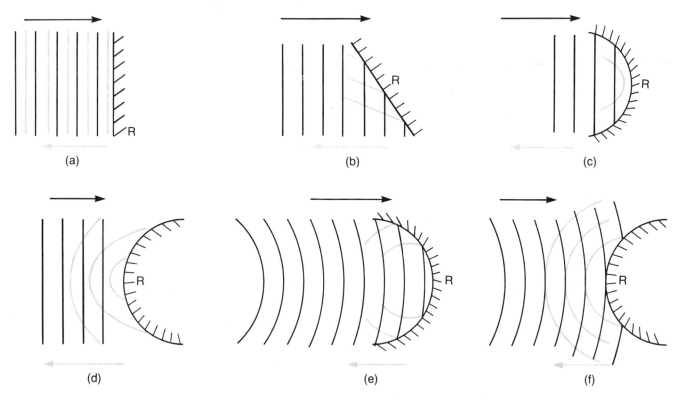

**Fig 14.5** Reflection of waves

In each case the behaviour of the waves after reflection can be worked out by considering which part of the incoming waves reaches the particular barrier first. This part of the wave will be reflected first. For example, in Fig. 14.5(d), the centre of each incoming straight wave obviously strikes the reflecting barrier first. The centre of each returning wave will therefore be in front, giving the curved shape shown.

In Fig. 14.5(c) and (e) the waves after reflection are approximately circular and converge to a point. This point is the image formed by waves from the point source (a distant source in Fig. 14.5(c)) after reflection at the barrier. In Fig. 14.5(d) the reflected waves appear to come from a point behind the barrier. This point represents a virtual image of the source.

## 14.4 Refraction

If water waves come to a region where their speed changes abruptly, then their direction of travel may change abruptly. The best way of achieving this change in speed is to alter the depth of water by placing a sheet of glass in the tray, since it is found that if the water is shallower the speed is reduced. This can be seen as waves approach a sloping beach. The body of the waves slows, but the crest is less affected and falls forward, causing the waves to break.

The sheet of glass should be placed in the tray so that it is covered by water to a depth of only one or two millimetres. A little experimentation will give the best depth. Straight waves, produced by either the beam or dowel rod, reach the leading edge of the glass plate at an angle

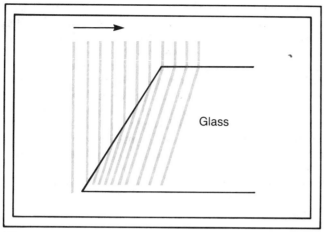

**Fig 14.6** Refraction of waves at a plane boundary

(Fig. 14.6). As soon as part of the wave comes over the shallow region it slows down. It is clear that the end of the wave which has been in the shallow region longer will fall behind the other end. Thus the direction of the wavefront changes as shown.

## 14.5 Interference

If water waves from two sources with the same frequency and similar amplitude meet, complete calm can be seen on some regions of the water surface. The waves are said to **interfere**.

Two spheres near the centre of the beam in the ripple tank are turned down so that when the beam vibrates two sets of circular waves are produced (Fig. 14.7). The interference pattern produced is shown in the same diagram.

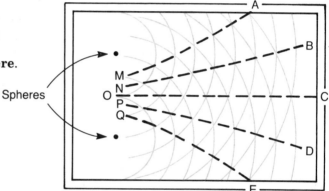

**Fig 14.7** Interference of waves

The two spheres vibrate in step and thus points along the line *OC*, which are all equidistant from the two spheres, will always receive the waves in step. Along this line the waves will always **reinforce** each other making the actual displacement greater than either wave taken separately (Fig. 14.8(a)). In Fig. 14.7 points on the lines *NB* and *PD* are a distance of one

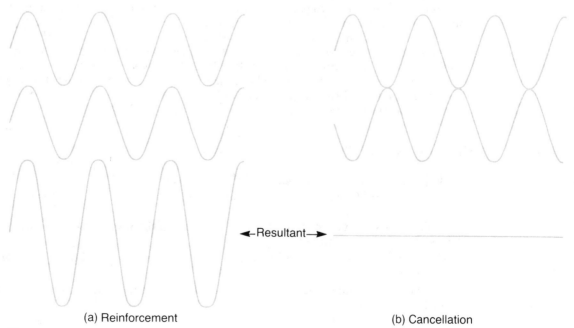

(a) Reinforcement  (b) Cancellation

**Fig 14.8** Addition of waves: (a) in step — reinforcement; (b) out of step — cancellation

wavelength further from one sphere than the other. Thus waves from the two spheres will arrive at each of these points one complete wavelength out of step and will again reinforce, as peaks arrive together. The same reasoning holds for lines *MA* and *QE*, except that the distance of each point from the two spheres differs by two wavelengths.

There will be directions between the lines mentioned along which points will be a distance of one, three, five, etc. half wavelengths further from one sphere than the other. Waves will arrive at these points either one, three or five (or any odd number) half wavelengths out of step. Thus complete **cancellation** of the two sets of waves will always occur at these points (Fig. 14.8(b)).

If the wavelength is reduced a smaller path difference is required for the waves from the two sources to be out of step by one, two, three, etc. complete wavelengths. Thus the lines in Fig. 14.7 come closer together and more regions of addition and cancellation of waves are seen in the tray. The reverse is true if the wavelength is increased.

If the two sources of waves are brought closer together the lines in Fig. 14.7 become further apart and fewer regions of addition and cancellation of waves are seen in the tray. If the sources are made further apart the reverse is true.

## 14.6 Diffraction

If straight water waves are passed through a narrow gap in a barrier or past a small object some bending of the waves round the edges of the barrier or the object is noticed. Thus some change in the direction of travel of the waves occurs round these edges. This effect is called **diffraction**.

For diffraction to be obvious the size of the gap or object has to be about the same as the wavelength of the waves. In Fig. 14.9(a) the gap between the barriers is much greater than the wavelength and little bending occurs. In Fig. 14.9(b) the gap size and the wavelength are about the same and the bending is marked. In fact the waves become circular after passing through the gap and look the same as if the gap were replaced by a point source of waves. In Fig. 14.9(c) where the object is about the same width as the wavelength the bending is again very noticeable.

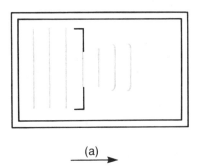 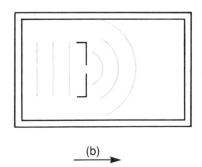 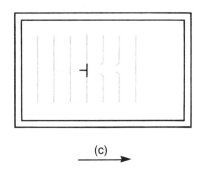

(a)  (b)  (c)

**Fig 14.9** Diffraction of waves: (a) wide gap; (b) narrow gap; (c) small object

The interference experiment described in the last unit may be carried out by allowing a series of straight waves to pass through two gaps in a barrier. After passing through the waves spread out in a semicircular fashion; that is they are diffracted, and produce the same interference pattern as was obtained with the two vibrating spheres. It should be noted that interference only occurs in this version of the experiment because diffraction takes place at the narrow gaps. If diffraction did not take place the waves would never meet.

A clear understanding of the behaviour of water waves discussed in this unit will make it easier to understand the properties of light and sound waves discussed in later units, as they behave in a similar way.

## 14.7 Summary

1. There are two types of waves
   (a) Longitudinal waves in which the motion causing the wave is in the same direction as the wave travels.
   (b) Transverse waves in which the motion causing the wave is at right angles to the direction in which the wave travels.
2. All types of wave motion, other than sound, are transverse. Sound is a longitudinal wave motion.
3. The equation connecting the velocity $v$, the frequency $f$ and the wavelength $\lambda$ of a wave motion is
$$\mathbf{v = f\lambda}$$

It is usual to measure the velocity in metres per second (m/s), the frequency in hertz (Hz) and the wavelength in metres (m).

4 Reflection is the change in direction of a wave motion when it hits a solid boundary.
5 Refraction is the change in direction of a wave motion caused by its change in speed.
6 Diffraction is the spreading out which occurs when waves pass round a small object or through a narrow gap.
7 Interference is the result of two wave motions meeting.

# 15 THE PASSAGE AND REFLECTION OF LIGHT

In some light experiments it is more convenient to consider the behaviour of **rays** rather than wavefronts. Rays are lines drawn at right angles to wavefronts and thus represent the direction in which the wave is travelling. A ray box is a device for producing rays; one version is shown in Fig. 15.1.

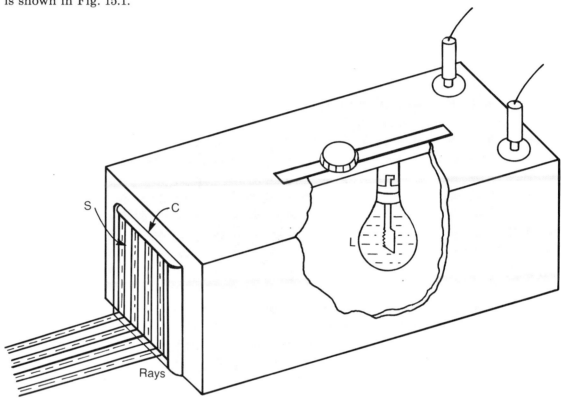

**Fig 15.1** Ray box

A small filament lamp $L$ is enclosed in a box with a cylindrical convex lens $C$ and a 'comb' $S$ containing parallel slits in front of it. A diverging, converging or parallel beam of rays (Fig. 15.2) may be obtained by moving the lamp $L$.

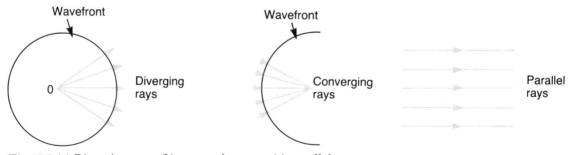

**Fig 15.2** (a) Diverging rays; (b) converging rays; (c) parallel rays

## 15.1 Rectilinear propagation

If a ray box is placed on a sheet of white paper and switched on, a ray is obtained through each slit of the comb. This ray is actually a very narrow beam of light with straight sharp edges. It can be said that light travels in straight lines (**rectilinear propagation**) in this case. The existence of shadows and eclipses of the sun and moon is further evidence that light travels in straight lines.

When an obstacle is placed in the path of light coming from a point source the shadow formed on a screen is uniformly dark and has sharp edges (Fig. 15.3). As no light reaches the region of shadow (umbra) it is concluded that light travels in straight lines.

# Rectilinear propagation

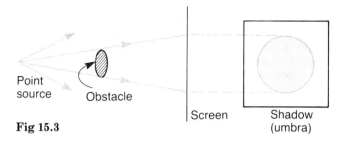

**Fig 15.3**

If an extended light source is used the shadow is edged with a border of partial shadow (penumbra). The area of partial shadow receives light from some points on the source, but other points on the source are obscured from it by the obstacle (Fig. 15.4).

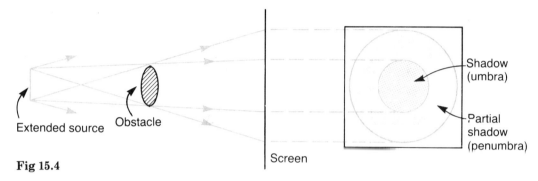

**Fig 15.4**

When the moon passes between the sun and the earth it casts a shadow or partial shadow on parts of the earth's surface (Fig. 15.5). This effect is known as an eclipse or partial eclipse of the sun. Area $c$ is total shadow, $b$ and $d$ are partial shadow, and $a$ and $e$ receive light from the whole of the sun's surface. On some occasions the moon is a little further from the earth than shown in the diagram and there is no area of complete shadow.

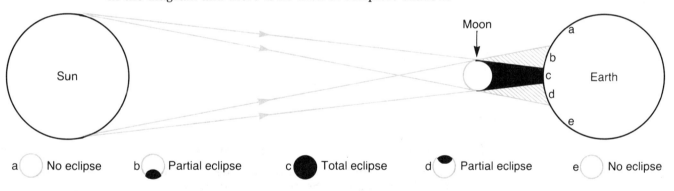

**Fig 15.5** Eclipse of the sun          Sun's appearance

The pin-hole camera relies on the fact that light travels in straight lines to produce a clear image. It is a very simple version of a camera, invented well before lenses were used to produce images (Fig. 15.6(a)).

A pin-hole camera may be made by removing the back of a small cardboard or metal box, and replacing it with a piece of semi-transparent paper (the screen). A pin-hole is punched in the side of the box opposite the screen. When the hole is held towards a bright lamp, such as a carbon filament lamp, in a darkened room, an inverted image of the lamp filament can be seen on the screen. A narrow beam of light from point $A$ on the source enters the camera through the pin-hole and strikes the screen at $A'$. Likewise light from $B$ arrives at $B'$. Narrow beams of light from all the different points on the object will fall on the screen between $A'$ and $B'$. Each point on the object is thus responsible for a point of light on the screen and a complete inverted

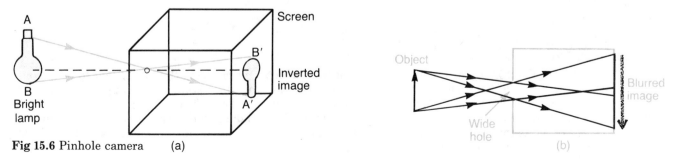

**Fig 15.6** Pinhole camera          (a)

image is seen. This image is formed on a screen and is said to be **real**. An image which cannot be shown on a screen, but only seen by the eye is called **virtual**.

If the pin-hole camera is moved closer to the lamp the image becomes bigger. The small hole means that little light enters the camera. A larger hole would improve this, but would lead to blurring of the image (Fig. 15.6(b)), unless a lens was used. Thus the exposure time needed to produce a picture on a film using a pin-hole camera is long.

## 15.2 Reflection at a plane surface

Reflection of light may be examined by using a plane mirror supported vertically on a sheet of white paper. A line $XN$, the normal, is drawn at right angles to the mirror surface so that $N$ is near the centre of the mirror (Fig. 15.7). Further lines are drawn from $N$ so that they are inclined at angles such as 20°, 30°, etc. to $XN$. A ray box is now placed so that a single ray follows one of the drawn lines. The position of the reflected ray is marked with dots. The ray box is moved to each of the lines in turn and the procedure repeated. In each case the angle of incidence $i$ of the ray is noted and the corresponding angle of reflection $r$ measured. Within the limits of experimental accuracy the two are found to be equal.

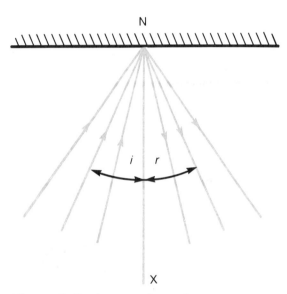

**Fig 15.7** Reflection at a plane mirror

The laws of reflection at plane surfaces are summarized as follows:

> **1 The incident ray, the reflected ray and the normal all lie in the same plane.**
> **2 The angle of incidence equals the angle of reflection.**

If a beam of divergent rays is reflected from a plane mirror, the rays will appear as in Fig. 15.8(a). A beam of convergent rays would give Fig. 15.8(b).

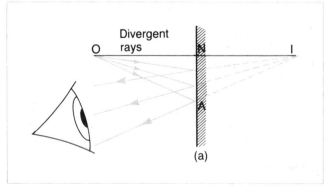

**Fig 15.8** (a) Diverging rays

In Fig. 15.8(a) the image $I$ is the point from which rays entering the eye appear to have come. These rays have been reflected according to the laws just stated and consideration of the triangles $ONA$ and $INA$ will show that they are similar. Thus the distance $IN$ is the same as the distance $ON$; that is the image is as far behind the surface of the mirror as the object is in front of it. As the rays do not actually come from $I$ the image is virtual. In Fig. 15.8(b) the reflected rays cross at the point $I$ to form a real image there.

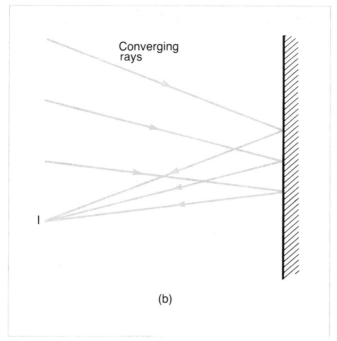

**Fig 15.8** (b) Converging rays

If a person looks at the image of an object in a mirror he notices that it appears the wrong way round; it is said to be **laterally inverted**. Figure 15.9 shows how this comes about. If $L$ and $R$ represent the left and right sides of the object viewed directly, it will be seen that $L'$, the image of $L$, appears on the right side of the image in the mirror.

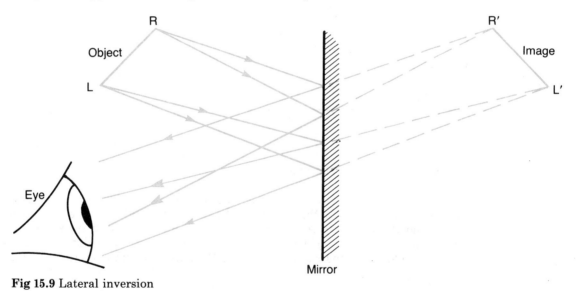

**Fig 15.9** Lateral inversion

## 15.3 Summary

1 Light normally travels in straight lines. This results in the formation of shadows and eclipses when an object is in a beam of light.
2 When a ray of light is reflected at a surface such as a mirror the angle of reflection equals the angle of incidence.

# 16 REFRACTION OF LIGHT

The laws of refraction are:

1 **The incident and refracted rays are on opposite sides of the normal at the point of incidence and all three are in the same plane.**
2 **The ratio of the sine of the angle of incidence to the sine of the angle of refraction is a constant.**

## 16.1 Refraction at a plane boundary

The refraction of rays of light can be studied using a ray box. We will consider what occurs when a single ray of light strikes a plane air to glass boundary (Fig. 16.1).

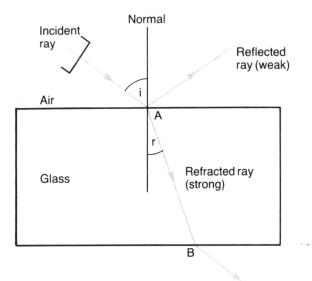

**Fig 16.1** Refraction at a plane boundary

The glass block stands on a sheet of white paper. A line at right angles to the boundary to be used is constructed near the middle of the face (a normal). It is arranged that the single ray strikes the glass block at $A$ at various angles of incidence, $i$, in turn. The position of this ray and the emerging ray in each case are marked on the paper with dots. The angle of refraction $r$, within the block, is constructed by joining the points $AB$ for each ray. The angle $r$ is measured

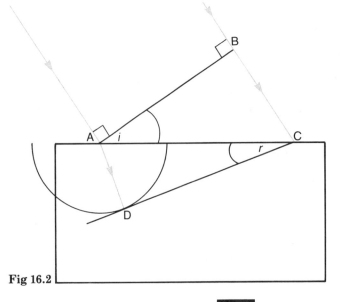

**Fig 16.2**

and, in each case, found to be less than the corresponding value of $i$; that is the ray always bends towards the normal. The ratio $\sin i/\sin r$ is calculated and found to be the same for all angles of incidence. The full significance of this result can be seen by studying Fig. 16.2.

$AB$ represents a plane wavefront striking the boundary at an angle $i$. While the end $B$ of the incoming wavefront is travelling the distance $BC$ to the boundary, the light which strikes $A$ will have entered the glass and will be passing through it at a lower speed. When $C$ is reached the light from $A$ must lie somewhere on the semi-circle shown. As the wavefront in the glass must be plane and also be passing through $C$ at this moment, it must be represented by the line $CD$.

Let the velocity of light in air be $v_1$ and in glass $v_2$. Also let $t$ be the time for the light to go from $B$ to $C$. It will also be the time for light to travel from $A$ to $D$.

Triangles $ABC$ and $ADC$ are both right angled triangles. Thus

$$\sin i = \frac{BC}{AC} \quad \text{and} \quad \sin r = \frac{AD}{AC}$$

Therefore $BC = AC \sin i$

and $AD = AC \sin r$

but $BC = v_1 t$

and $AD = v_2 t$ (distance = velocity × time)

thus $v_1 t = AC \sin i$

and $v_2 t = AC \sin r$

Dividing gives $\dfrac{v_1}{v_2} = \dfrac{\sin i}{\sin r}$

The ratio $\dfrac{v_1}{v_2}$ is known as the **refractive index $\mu$** between the two materials.

The velocity in a vacuum is the same for all electromagnetic waves. The velocity in any medium is less than in a vacuum and varies from one colour to another. Hence the refractive index of a medium is different for different colours.

Thus the ratio $\sin i/\sin r$ which the experiment showed to be constant is the ratio of the velocity of the light in the two materials. This is perhaps not surprising as it is the change in the velocity of light which leads to the refraction.

Figure 16.3 shows two consequences of the refraction which occurs when light passes from water to air and thus speeds up. In Fig. 16.3(a) the bottom of the pond appears to be at $I$ and not $O$. The pond appears more shallow than it is. In Fig. 16.3(b) the stick appears bent at the point where it enters the water.

The explanation is the same for both effects. We will first consider the case when the bottom of the pond or the end of the stick are viewed from vertically above (Fig. 16.3(c)). The ray $OBC$ in the diagrams is thus close to the normal to the water surface. The emergent ray $BC$ appears to be coming from a virtual image $I$, so that $AI$ is the apparent depth of the pond or end of the stick.

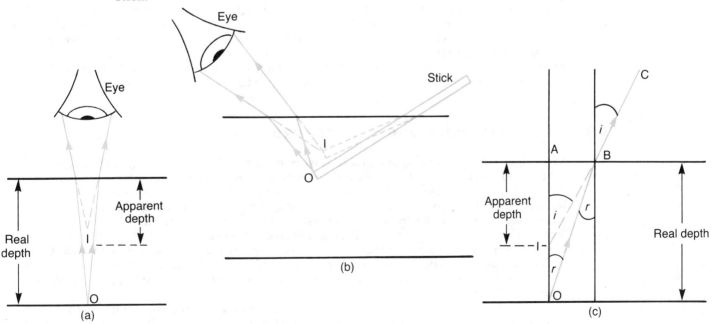

**Fig 16.3** Consequences of refraction: (a) apparent depth; (b) bent stick

The refractive index $\mu$ of water is given by:

$$\mu = \frac{\sin i}{\sin r}$$

but $\angle AIB = i$ and $\angle AOB = r$

therefore
$$\mu = \frac{\sin \angle AIB}{\sin \angle AOB}$$
$$= \frac{AB}{BI} \bigg/ \frac{AB}{BO} = \frac{BO}{BI}$$
$$= \frac{AO}{AI}$$

when $B$ is close to the normal at $A$, which is the case when viewing is done from directly above $O$. Thus

$$\mu = \frac{\text{Real depth}}{\text{Apparent depth}}$$

If the viewing is done more obliquely the apparent reduction in depth is more but this formula is no longer valid.

Another example of refraction at a plane surface takes place in a prism (Fig. 16.4).

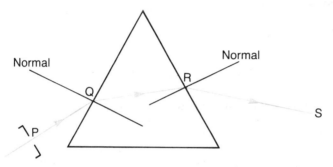

**Fig 16.4** Refraction by a prism

When light enters the prism at $Q$, its velocity is reduced and refraction takes place towards the normal. The reverse takes place at $R$ as the light leaves the prism. The prism is a glass block. In the case of a rectangular glass block, the refracting faces are parallel, and the emerging ray is parallel to the incident ray. For a prism the refracting faces are not parallel and this results in the emerging ray not being parallel to the incident ray. The angle between these two rays is called **the angle of deviation**. If the incident light is not all of the same wavelength, but a mixture, different wavelengths will undergo slightly different deviation at each face. Thus, for example, blue light will be deviated through a greater angle than red light.

A spectrum is obtained if a parallel beam of white light falls on a prism in the direction $PQ$ (Fig. 16.4) and a screen is placed on the far side of the prism. The separate colours are visible in order from red to violet, red being the least deviated. However, the spectrum is not very good: a pure spectrum is obtained if a converging lens is added on the far side of the prism such that the screen passes through the focal point of the lens. This lens then focusses the parallel rays of the different colours which leave the prism; each colour to one place on the screen.

## 16.2 Internal reflection and critical angle

When light passes from one medium to a more optically dense medium (i.e. the speed of the light is reduced) refraction occurs for all angles of incidence, together with a very small amount of reflection. But refraction does not always occur at the surface of a less optically dense medium, for example when light is passing from glass or water to air.

Consider a ray passing from glass or water to air with a small angle of incidence (Fig. 16.5(a)). Here we get both a refracted and a reflected ray, the latter being relatively weak. If the angle of incidence is gradually increased the reflected ray becomes stronger and the refracted ray weaker, until for a certain **critical angle of incidence $c$**, the angle of refraction is just $90°$. For angles of incidence greater than $c$ the ray of light will not pass through. This special case is shown in Fig. 16.5(b).

Since it is impossible to have an angle of refraction greater than $90°$, it follows that all the light is internally reflected for angles of incidence greater than the critical angle. There is no refracted ray and this condition is known as **total internal reflection** (Fig. 16.5(c)).

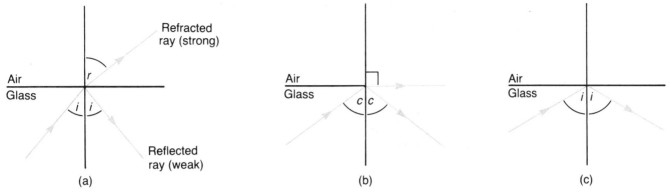

**Fig 16.5** (a) Refraction; (b) critical angle; (c) total internal reflection

$$\text{Now} \quad \frac{\sin i}{\sin r} = \frac{v_2}{v_1} = \frac{1}{\mu}$$

For critical angle $i = c$ and $r = 90°$. Hence $\sin r = 1$. Thus

$$\frac{v_2}{v_1} = \sin c$$

The ratio of the velocity of light in glass or water to its velocity in air (greater) is $v_2/v_1$ in this case.

The value of the critical angle for a glass to air boundary is about 42° and for water to air about 48°.

## 16.3 Refraction at a spherical boundary

Refraction of light at a spherical surface follows the same rules as refraction at a plane surface; that is:

1 The incident and refracted rays are on opposite sides of the normal at the point of incidence and all these are in the same plane.
2 The ratio of the sine of the angle of incidence to the sine of the angle of refraction is a constant.

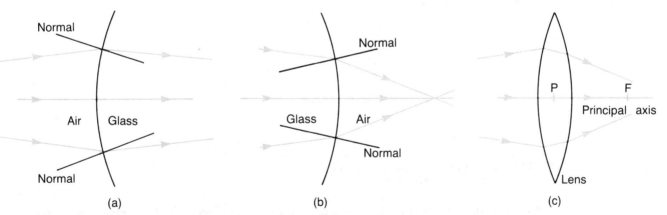

**Fig 16.6** Refraction at lens surfaces

Consider Fig. 16.6(a) showing three rays passing from air to glass. The three rays are all refracted according to the rules just stated; however, in this case, refraction results in the rays being brought together or focussed. If the rays leave the glass, through a second boundary curved the other way, before they have come together, further focussing occurs (Fig. 16.6(b)), as they bend away from the normal. The combined effect of the two surfaces is to provide us with a converging lens as shown in Fig. 16.6(c).

The point **F** is the point through which rays incident parallel to the principal axis pass after refraction by the lens. It is called the **focal point** or **principal focus** of the lens. The distance **PF** is known as the **focal length** $f$.

The power of the lens is defined by

$$\text{power} = \frac{1}{f}$$

and if $f$ is measured in metres the unit of power is called a dioptre.

The behaviour of a diverging lens is shown in Fig. 16.7.

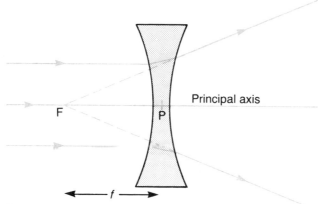

**Fig 16.7** Diverging lens

The point **F** is the point from which rays incident parallel to the principal axis appear to come after refraction by the lens. It is called the **focal point** of the lens. The distance **PF** is known as the **focal length** $f$. Because the rays do not actually pass through the focal point $F$ it is virtual and the focal length $f$ is negative therefore. The power is calculated in the same way as for a converging lens, but because $f$ is negative the power is also negative.

## 16.4 To determine the focal length of a converging (convex) lens

The focal length may be measured accurately by placing a plane mirror behind the lens and an illuminated point object in front of it (Fig. 16.8). The object $O$ is moved until a clear image $I$ of it is obtained alongside, due to reflection at the plane mirror. The distance $PO$ is the focal length $f$.

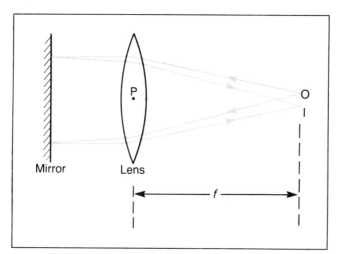

**Fig 16.8** Plane mirror method for determining the focal length of a convex lens

As the image occurs almost at the same point as the object, the light must return from the mirror almost along its incident path. This means that the light strikes the mirror normally; the light incident on the lens from the mirror is thus parallel with the principal axis.

An approximate value for the focal length of a converging lens may be found by casting the image of a distant object, such as the laboratory window in strong daylight, on a screen and measuring the distance between the lens and the screen.

## 16.5 Construction of ray diagrams

An image of a point on an object is formed where two rays from that point intersect after refraction. When constructing diagrams for converging lenses it is usually simplest to use any two of the following three rays:

1 A ray parallel to the principal axis which passes through the focal point after refraction.
2 A ray through the focal point which emerges parallel to the principal axis after refraction.
3 A ray through the centre of the lens which is undeviated.

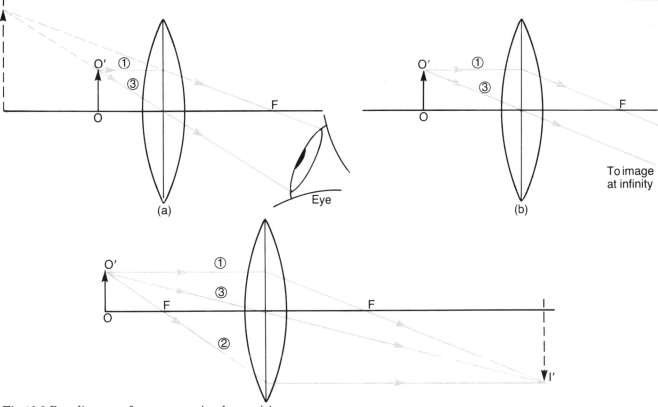

**Fig 16.9** Ray diagrams for a converging lens

Ray diagrams are most conveniently drawn to scale on squared paper. The height chosen for the object does not matter as the height of the image will always be in proportion. Figure 16.9 gives a series of diagrams to show the type of image formed for different distances of an object from a converging lens. In all cases the magnification is given by:

$$\text{magnification} = \frac{\text{height of image}}{\text{height of object}} = \frac{II'}{OO'}$$

$$\text{magnification} = \frac{\text{image distance}}{\text{object distance}}$$

When constructing ray diagrams for diverging lenses it is usually simplest to use any two of the following three rays:

1. A ray parallel to the principal axis which appears to come from the focal point $F$ after refraction.
2. A ray incident in the direction of the focal point which emerges parallel to the principal axis after refraction.
3. A ray through the centre of the lens which is undeviated.

Figure 16.10 shows the formation of an image by a diverging lens.

For all positions of the object the image is virtual, erect and smaller than the object, and is situated between the object and the lens.

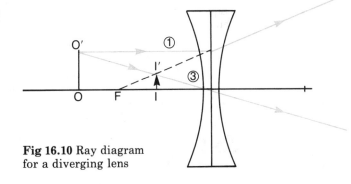

**Fig 16.10** Ray diagram for a diverging lens

For both converging and diverging lenses it can be shown that:

$$\frac{1}{u} + \frac{1}{v} = \frac{1}{f}$$

where  $u$ = object distance from lens
  $v$ = image distance from lens
  $f$ = focal length of lens

Virtual distances carry a negative sign, including the focal lengths of diverging lenses.

## 16.6 Simple microscope (magnifying glass)

Figure 16.9(a) illustrates the use of a lens as a magnifying glass. Suppose that, without the lens, $I$ is the closest to the eye that the object could normally be placed for clear focussing. Using the lens the object may now be placed at $O$ and still be seen clearly as, to the eye, the object appears to be at $I$. The lens has thus enabled the object to be placed closer to the eye than would be the case without it, and the object appears larger because it appears to be closer.

## 16.7 The camera

The camera consists of a lens and sensitive film mounted in a light-tight box. The distance of the lens from the film may be adjusted using the focussing ring. This is marked in metres to show the distance at which an object will give a clear image. The amount of light entering the camera is controlled by a diaphragm of variable aperture (hole size) and the speed of the shutter. These used together allow the correct amount of light to reach the film and give the right exposure. The main features are shown in Fig. 16.11.

The ray diagram for a camera is shown in Fig. 16.9(c).

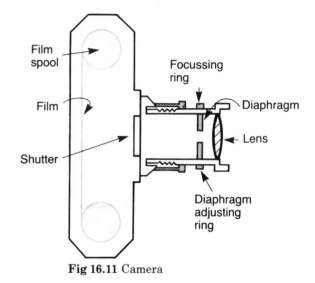

**Fig 16.11** Camera

## 16.8 The eye

A simplified diagram of the human eye is shown in Fig. 16.12. In many ways the eye is similar to the camera. An image is formed by the lens on the sensitive retina at the back of the eye.

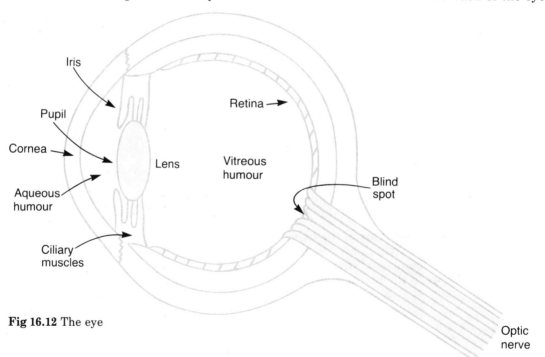

**Fig 16.12** The eye

The iris automatically adjusts the size of the circular opening in its centre, known as the pupil, according to the brightness of the light falling on it. Focussing of the image on the retina is achieved partly by refraction at the curved surface as light enters the eye, and partly by the action of the lens. The ciliary muscles vary the thickness of the lens and hence its focal length. When the muscles are relaxed the lens is thin. Unlike the camera, the position of the lens does not alter.

If the ciliary muscles weaken, the eye lens cannot be made sufficiently fat to clearly focus close objects. However, distant objects can be clearly focussed. The eye is said to suffer from **long sight**. The same defect occurs if the eyeball is too short, and it may be corrected by using spectacles containing a converging lens (Fig. 16.13(a)). In the case of **short sight** the muscles do not relax sufficiently and consequently distant objects are focussed in front of the retina. Short sight also occurs if the eyeball is too long. The defect can be corrected with a diverging lens (Fig. 16.13(b)).

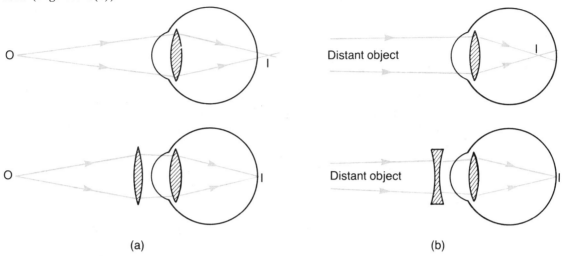

(a)           (b)

**Fig 16.13** Eye defects and their correction: (a) long sight; (b) short sight

In the case of long sight bifocal spectacles are frequently worn. Each eyepiece is in two halves, the upper half often being a plane sheet of glass through which to look at distant objects and the lower half being a converging lens through which to view close objects, such as a book or newspaper. Such an arrangement removes the necessity to be continually removing and replacing spectacles.

## 16.9 Summary

1. When light enters a different material its speed changes. This results in a change in its direction of motion (refraction) unless it crosses the boundary between the two materials at 90°. For example, when a ray of light enters glass from air it bends towards the normal. On leaving it bends away from the normal.
2. The refractive index of a material is the ratio of the speed of light in a vacuum to its speed in the material.
3. When looking at an object in a material at 90° to the boundary with air the following relationship holds:

$$\text{Refractive index of the material} = \frac{\text{real depth}}{\text{apparent depth}}$$

4. The focal point $F$ of a lens is the point through which rays incident parallel to the principal axis pass after refraction by the lens. The distance between $F$ and the centre of the lens is known as the focal length $f$.
5. The power of a lens $= \frac{1}{f}$ where $f$ is in metres and the power in dioptres.
6. A converging lens bends rays of light inwards (towards the principal axis). Its focal length is positive. A diverging lens bends them away. Its focal length is negative.
7. For any lens

$$\frac{1}{u} + \frac{1}{v} = \frac{1}{f}$$

where $u$ is the distance the object is away from the lens and $v$ the distance that the image is away from the lens. The distances of real objects and images are treated as positive, those of virtual objects and images as negative.

8. Long sight occurs if the eye lens cannot be made sufficiently fat (strong) to clearly focus close objects or the eyeball is too short.
9. Short sight occurs if the eye lens cannot be made sufficiently thin (weak) to clearly focus distance objects or the eyeball is too long.

# 17 THE WAVE BEHAVIOUR OF LIGHT

## 17.1 Diffraction

The **diffraction** or spreading out which occurs when waves pass through a narrow gap or round a narrow object has already been mentioned in the unit on water waves. It was then stated that the size of the gap or object had to be about the same as the wavelength of the waves, if significant bending was to occur. Light has a wavelength (about $5 \times 10^{-7}$ m) very much less than the wavelength of water waves. The dimensions of the gap or object must be very small therefore if diffraction of light is to be observed. In practice a slit which is sufficiently narrow for the purpose can be made by drawing a sharp razor blade across the surface of a blackened microscope slide.

If the slide is held close to the eye and a light filament viewed through the slit, the filament looks much broader than it does with the naked eye, thus showing that light has been diffracted round the edges of the slit. The effect is most obvious if the filament and slit are parallel to each other. A similar effect can be obtained by partially closing one's eyes, until there is a narrow slit between the eyelids, and then looking at a light.

## 17.2 Interference

Figure 17.1 shows a simple arrangement (Young's slits) to demonstrate the **interference** of light.

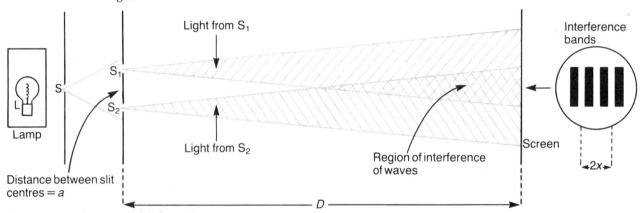

**Fig 17.1** Interference of light waves — Young's experiment

$S_1$ and $S_2$ are two slits marked on a microscope slide as suggested above. If interference between light passing through the two slits is to be observed they should be 0.5 mm or less apart. They are illuminated by bright light from a lamp placed behind the single slit $S$; the purpose of the single slit being to cut out as much stray light as possible. The interference effect, in the form of light and dark fringes (Fig. 17.1), is observed on a screen placed a metre or so from the slits.

At certain places on the screen the light waves from the two slits will be in step (Fig. 14.8(a), p. 53), and bright fringes are seen. This occurs for places on the screen which are one, two, three, etc. wavelengths further from one slit than the other, or where there is no path difference (central bright fringe). These places correspond to the directions in Fig. 14.7 where the water waves reinforce.

Waves from the slits arriving at places on the screen about midway between the bright fringes will be completely out of step. These places are one, three, five, etc. half wavelengths further from one slit than the other, and correspond to the directions of calm water in Fig. 14.7.

If the slits in Fig. 17.1 are drawn closer together the fringes on the screen are more spread out and vice versa. If the lamp used produces white light, which is a mixture of wavelengths, fringes for the different colours will occur at slightly different positions on the screen. The centre fringe will be white but colours will be seen after two or three fringes to either side of this.

The equation $x = \dfrac{\lambda D}{a}$ connects the distances shown in Fig. 17.1, where $x$ is the fringe separation, $D$ the distance between the plane of the slits and the screen, $a$ the separation of the slits and $\lambda$ the wavelength of the light used.

## 17.3 The diffraction grating

A **diffraction grating** consists of a large number of parallel lines ruled very close together (between 500 and 3000 lines per centimetre) on a transparent plate; it is usually produced photographically. Light passing through each of the narrow gaps between the rulings is diffracted and overlaps in the region beyond the grating. The grating thus acts like a multi-light source. An arrangement using a diffraction grating is shown in Fig. 17.2.

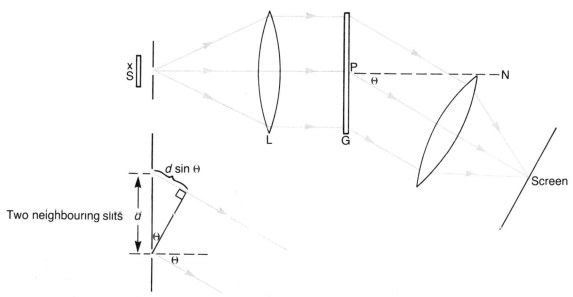

**Fig 17.2** Diffraction grating

A bright lamp $S$, with a colour filter in front, is placed behind a narrow slit. A converging lens $L$ is positioned with the slit at its focus, so that a parallel beam of light is incident normally on the grating $G$. The emerging parallel beam in a given direction is then collected by moving a lens round from the direction $PN$ until an image of the slit is first received on the screen. The image occurs at an angle $\theta$ such that the path difference between light rays collected from neighbouring slits is one wavelength. A second image is received when the path difference is two wavelengths etc.

The path difference between rays from neighbouring slits is given by $d \sin \theta$, where $d$ is the distance between the centres of such slits. Thus when $d \sin \theta = n\lambda$ where $n = 1, 2, 3$, etc., reinforcement occurs and an image is received. The values of $d$ and $\theta$ can be measured and a value obtained for the wavelength $\lambda$.

The filter is used to produce light of approximately one particular wavelength. If white light is used images occur at different angles for different wavelengths and a series of spectra are received. The central image is white.

The apparatus may be simplified by dispensing with the two lenses. If the lamp is placed a long way in front of the grating the light incident on the grating will be parallel without using a lens to make it so. If the images are viewed directly by eye rather than on a screen the eye lens replaces the second lens.

## 17.4 The electromagnetic spectrum

The light waves discussed in recent pages are just one small part of a family of waves called the **electromagnetic spectrum**. This family includes radio and television waves, infrared rays, light, ultraviolet light, X-rays and $\gamma$-rays. All members travel with a velocity of $3 \times 10^8$ m/s in space. Their difference lies in their frequency and wavelength. Radio waves lie at the long wavelength–low frequency end of the spectrum and $\gamma$-rays at the short wavelength–high frequency end. Because of their differing frequencies and wavelengths, different regions of the spectrum exhibit different properties. Some of these properties are summarized in Table 17.1.

Radio waves form the basis of all long distance 'wireless' communication. The behaviour of these waves is determined largely by the presence round the earth of the ionosphere. This region of ionized gas, at heights between 80 and 400 km above the earth's surface, acts as a mirror for radio waves of many frequencies. Waves of long wavelength and low frequency are reflected by this layer and such waves are thus very useful for communications round the earth's surface. Waves of short wavelength and high frequency (>30 MHz) are able to penetrate the ionosphere and are used for all communications with artificial satellites. Radio waves can pass comparatively easily through brickwork and concrete.

## Table 17.1

| Wavelength ($\lambda(m)$) | Type | Production | Reflection | Refraction | Diffraction and interference |
|---|---|---|---|---|---|
| $>10^{-4}$ | Radio | Electrons oscillating in wires | Ionosphere | Atmosphere | Two stations |
| $7 \times 10^{-7} \to 10^{-4}$ | Infrared | Hot objects | Metal sheet | | |
| $4 \times 10^{-7} \to 7 \times 10^{-7}$ | Visible | Very hot objects | Metal sheet | Glass | Grating |
| $10^{-9} \to 4 \times 10^{-7}$ | Ultraviolet | Arcs and gas discharges | | | |
| $10^{-12} \to 10^{-8}$ | X-rays | Electrons hitting metal targets | | | Crystals |
| $<10^{-10}$ | $\gamma$-rays | Radioactive nuclei | | | |

Any object heated nearly to red heat is a convenient source of infrared radiation. The radiation is readily absorbed (and also emitted) by objects with rough black surfaces, but strongly reflected by light polished ones. This radiation may therefore readily be detected by painting the bulb of a thermometer black and allowing the radiation to fall on the bulb. The thermometer reading rises.

Radiation with wavelengths just shorter than that of the violet end of the visible spectrum is termed ultraviolet radiation. It is emitted by any white hot body, such as the filament of an electric light bulb. However, in this case the radiation is absorbed by the glass envelope of the bulb. A discharge tube containing mercury vapour (with quartz envelope) is a more intense source of ultraviolet radiation. A great deal of ultraviolet radiation is emitted by the sun, but the majority of this is absorbed by the earth's atmosphere and only a small fraction reaches the surface of the earth. It is this which causes browning of the skin. One of the best known properties of ultraviolet radiation is its ability to cause substances to **fluoresce**. This is the term given to the emission of visible light by substances when ultraviolet radiation is shone on them. Examples are real diamonds, uranyl salts and paraffin oil.

The X-ray region of the electromagnetic spectrum is defined more by the method of generation of the radiation than by its precise wavelength. The method used is to cause electrons to hit a metal target. X-rays are used in medicine. Their ability to penetrate matter depends on the atomic number (see Unit 25.2) of the nuclei of the material through which they pass. X-rays are comparatively easily absorbed by bone which contains calcium (Z = 20), whereas they pass much more readily through organic material which contains hydrogen (Z = 1) and carbon (Z = 6). In addition, X-rays affect a photographic plate in a similar manner to visible light. An X-ray photograph may therefore be taken of the human leg, for example, and a bone fracture detected. X-rays are dangerous in too intense quantities. They eject electrons from the region on which they fall and can cause damage to living cells. In controlled quantities they can be used to kill off diseased tissue. X-rays like other forms of electromagnetic radiation show diffraction effects. The atomic spacing in crystals is comparable with the wavelength of X-rays, and thus X-rays are appreciably diffracted by atoms in crystals. The development of the science of X-ray crystallography has led to a much more detailed understanding of the structure of materials.

$\gamma$-rays are distinguished from X-rays in that they are emitted by the nuclei of natural or artificial radioactive materials. They are generally more penetrating, more dangerous and more difficult to screen than X-rays. They are mentioned in more detail in the unit on radioactivity (Unit 25.4).

## 17.5 Summary

1. Diffraction is the spreading out which occurs when light waves pass through a narrow gap or round a narrow object. The width of the gap or object should be equal to or smaller than the wavelength of the light if spreading out is to be significant.
2. When two sets of waves meet interference occurs. Where two crests or two troughs meet a large crest or trough results. Light is seen in these regions. Where a crest and a trough meet cancellation occurs and no light is seen. Such regions are dark.
3. In Young's experiment light from one lamp passes through two narrow slits which are close together. The diffraction which occurs at the slits results in the light overlapping beyond the slits and interference occurs.
4. A diffraction grating acts in a similar way to the two slits in Young's experiment. There are very many slits on the grating (between 500 and 3000 per centimetre) and they are very narrow.

# 18 SOUND

**Sound waves** differ from waves of the electromagnetic spectrum in that they are mechanical waves requiring a medium through which to pass. This may be demonstrated by placing an electric bell inside a glass jar from which the air can be removed (Fig. 18.1).

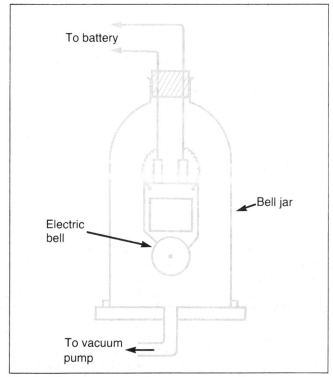

**Fig 18.1** Passage of sound

When all the air has been pumped from the jar the ringing can no longer be heard, although the hammer can still be seen striking the gong. The faint vibration which is still audible comes from the passage of sound through the connecting wires.

Sound waves travel through solids and liquids as well as gases. If one person places his ear near a long metal fence while a second person gives the fence a tap some distance away two sounds will be heard. The first is due to transmission through the fence; the second comes through the air. Sound travels about 15 times faster through a solid than through air. Fishermen are using transistor 'bleepers' which, when lowered into the sea, attract fish up to one mile away. The speed of sound in water is about 1400 m/s.

Sound waves are **longitudinal** rather than transverse; that is the wave particles oscillate in the same direction as the wave travels and not at right angles to it. Owing to their longitudinal nature sound waves consist of a series of **compressions** and **rarefactions**.

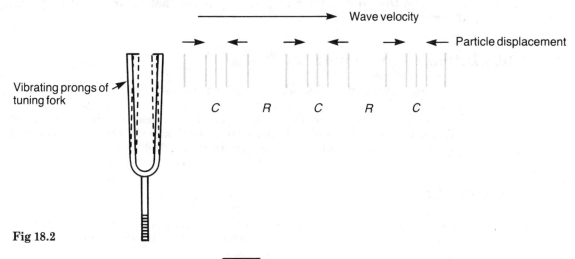

**Fig 18.2**

Figure 18.2 shows how a vibrating tuning fork sends out a sound wave. The prongs are set vibrating by striking the fork. When the right-hand prong moves to the right it pushes the layers of air in that direction to the right. These layers are thus pushed closer together; that is, they are compressed. This disturbance is transmitted from one layer of particles to the next with the result that a compression pulse or high pressure region moves away from the fork. Similarly, when the prong moves to the left, a low pressure region or rarefaction occurs to its right. Compressions and rarefactions move alternately through the air. The particle at the centre of a compression is moving through its rest position in the same direction as the wave, whilst a particle at the centre of a rarefaction is moving through its rest position in the opposite direction to the wave. At regular intervals in the material through which the sound is travelling, there will be particles undergoing exactly the same movement at the same moment. Such particles are said to be **in phase**.

As in the case of transverse waves, the distance between two successive particles in the same phase is called the **wavelength** $\lambda$.

The **amplitude** $a$ of a wave is the maximum displacement of a particle from its rest position.

The **frequency** $f$ is the number of complete oscillations made in one second. The unit is the hertz (Hz).

The **velocity** $v$ is the distance moved by the wavefront in one second.

As in the case of transverse waves, the velocity, frequency and wavelength are related by the equation:

$$v = f\lambda$$

which is proved in the unit on progressive waves (see Unit 14.1).

## 18.1 Echoes

Echoes are produced by the reflection of sound from a hard surface such as a wall or cliff. Sound obeys the same laws of reflection as light. Thus if a person claps his hands, when standing some distance from a high wall, he will hear a reflection some time later. In order that the reflection may be heard separately from the original clap it must arrive at least 1/10 s later. As the speed of sound in air is 340 m/s, this means that the wall must be at least 17 m away.

Echoes may be used to measure the speed of sound in air reasonably accurately. One person should make the sound and another carry out the timing. The first person claps his hands and listens for the echo from a wall. For accurate results the time interval will need to be about half a second; thus he needs to be about 100 m from the wall. Having obtained an estimate of the time interval, he continues to clap his hands and adjusts the rate of striking until each clap coincides in time with the arrival of the echo of the previous clap. When the correct rate of striking has been achieved the second person times 30 or more intervals between claps with a stopwatch. The speed of sound is then calculated by dividing the distance to the wall and back by the interval between two successive claps. The experiment should be repeated several times and an average value calculated.

## 18.2 Pitch

The **pitch** of a note depends on its frequency relative to other notes. This can be demonstrated by connecting a signal generator to a loudspeaker. If the frequencies 120, 150, 180 and 240 Hz (that is a ratio of 4:5:6:8) are produced in quick succession, the familiar sequence doh, me, soh, doh will be recognized. However, the same familiar sequence is noted when the frequencies 240, 300, 360 and 480 Hz are produced one after the other. The musical relation between notes thus depends on the ratio of their frequencies rather than their actual frequencies. An octave is always in the ratio 1:2.

## 18.3 Intensity and loudness

The most important factor affecting the **intensity** of a sound of given frequency is its **amplitude**. The intensity depends on the square of the amplitude and thus if the amplitude is doubled the intensity increases by a factor of four. Intensity is a measure of energy.

The **loudness** of a sound will obviously depend on its intensity. However, it also depends on the sensitivity of the human ear to sounds of different frequency.

## 18.4 Quality

A piano note can be distinguished from a trumpet note of the same pitch and loudness. The property which distinguishes them is known as **quality** or **timbre**. Musical instruments emit not only the basic or **fundamental** note but also **overtones**. An overtone (or harmonic) is a multiple of the fundamental frequency. Different instruments emit varying quantities of different overtones and it is the quantity of each present which determines the quality of the note. The note from the trumpet possesses a quality derived from the presence of strong overtones of high frequency.

## 18.5 Summary

1. Sound is a longitudinal wave motion.
2. The equation $v = f\lambda$ applies (see Unit 14.1).
3. Sound is reflected according to the same rules as light. Reflected sound is often called an echo.
4. The pitch of sound depends on its frequency.
5. The intensity of sound is a measure of energy and depends on the square of the amplitude of the sound wave.

# 19 MAGNETS

Certain materials have the property to attract iron; this property is known as **magnetism**. One such material is an iron ore called lodestone or magnetite; others are iron, steel, cobalt and nickel. Recently various alloys have been produced which can be made into very strong magnets. Alni, alcomax and ticonal are alloys of nickel and cobalt which are used for making powerful permanent magnets. Mumetal is an alloy which has been developed for the electromagnet and transformer, in which temporary magnets are used.

When a bar magnet is placed on a cork floating on water, so that it can swing in a horizontal plane, it comes to rest with its axis approximately in the north–south direction. The vertical plane in which the magnet lies is called the **magnetic meridian**. The end which points towards the north is called the north-seeking pole, or the N pole for short, and the other end the south-seeking or S pole.

If the N pole of a second magnet is brought near the N pole of the magnet on the floating cork, repulsion occurs and the cork and magnet tend to swing round. Repulsion occurs between two S poles in a similar way. However, a N and a S pole attract one another. These results may be summed up by saying:

**Like poles repel, unlike poles attract**

## 19.1 Making magnets

If a piece of iron is placed near a magnet it will become a magnet, but when it is removed it loses its magnetism. There are, however, several methods of making a more permanent magnet.

Magnets were originally made by stroking steel bars with a lodestone or another magnet. This was done in one of two ways, both illustrated in Fig. 19.1.

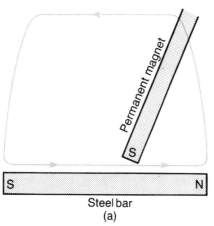 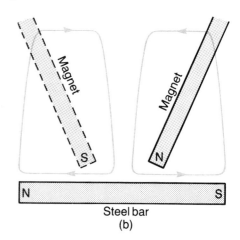

**Fig 19.1** (a) Single touch; (b) double touch

The single touch method requires one magnet (Fig. 19.1(a)). One pole of a magnet is stroked along a steel bar from one end to the other repeatedly. After a sufficient number of strokes the bar will begin to behave as a magnet. One of the poles produced is concentrated in one part of

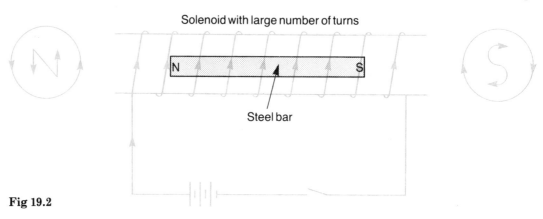

**Fig 19.2**

the bar where the stroking ends, but the other pole is apt to be spread along the bar where the stroking starts and this is a disadvantage of this method. In order to overcome the problem and concentrate the poles at both ends the method of double touch is used. The steel is stroked from the centre outwards with the unlike poles of two magnets simultaneously (Fig. 19.1(b)). In both methods the polarity produced at the end of the bar where stroking ceases is opposite to that of the stroking pole.

Magnetization by an electrical method (Fig. 19.2) is a far quicker and more efficient method of making magnets than any of those already described. A cylindrical coil is wound with about 500 turns of insulated copper wire and connected to a direct current supply. A coil of this kind is called a **solenoid**. A steel bar is placed inside the coil and the current switched on. The current creates a strong magnetic field within the coil and the steel immediately becomes a magnet, retaining its magnetism when the current is switched off.

## 19.2 Magnetic fields

In the space around a magnet a force is exerted on a piece of iron. This region is called a **magnetic field**. A magnetic field may be plotted by using a plotting compass to follow the lines of force. Figure 19.3 shows how this is done.

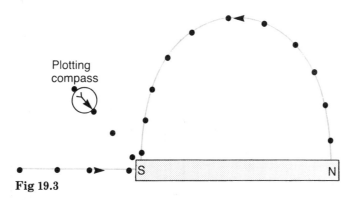

**Fig 19.3**

A bar magnet is placed on a sheet of white paper. Starting near one end of the magnet, the positions of the ends of the compass needle are marked by pencil dots. The compass is then moved until the near end of the needle is exactly over the dot furthest from the magnet and a third dot made under the other end of the needle. This process is repeated many times until the compass reaches the other end of the magnet. Further lines of force may then be plotted in a similar way. Conventionally the lines of force are labelled with an arrow indicating the direction in which a north pole would move. The strength of the magnetic field in a particular region is indicated by the closeness of the lines of force.

A plotting compass is sensitive and can be used to plot relatively weak fields. It is unsuitable for fields in which the direction of the lines of force changes rapidly in a short distance, for

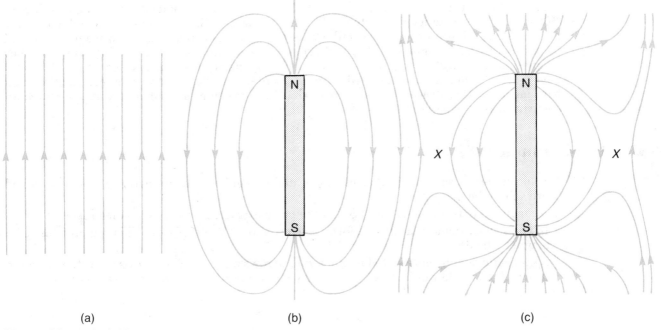

**Fig 19.4** Magnetic fields

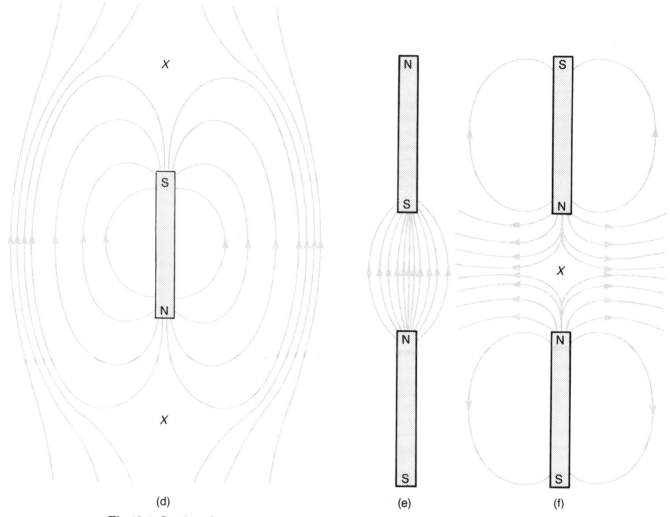

(d)  (e)  (f)

**Fig 19.4** *Continued*

example in the region of two magnets placed close together. These fields are best investigated using iron filings, although these do not indicate in which of the two possible directions the magnetic field is acting.

The magnets whose field is to be studied are placed beneath a sheet of stiff white paper. A thin layer of iron filings is then sprinkled from a caster. If the paper is tapped gently the filings form a pattern indicating the lines of force. Each filing becomes magnetized by induction and when the paper is tapped the filings vibrate and are able to turn in the direction of the magnetic force.

Figure 19.4 illustrates the magnetic field close to various arrangements of magnets. Figure 19.4(a) shows the magnetic field due to the earth alone, while (b) shows the field due to a bar magnet alone. Figure 19.4(c) and (d) illustrate the fields resulting when a bar magnet is placed in the earth's field. Figures 19.4(e) and (f) show the magnetic field resulting when two bar magnets are placed close together end to end. The points marked $X$ are regions where the total field is zero. Such regions are called **neutral points**. The shape of the magnetic field round a bar magnet — Fig. 19.4(b) — is similar to that in the region of a solenoid through which an electric current is passing.

## 19.3 Summary

1 Some materials such as iron, steel, cobalt and nickel have the property to attract iron and are said to be magnetic.
2 A magnet has two poles, N and S. Unlike poles attract, whereas like poles repel.
3 In the space around a magnet a force is exerted on a piece of iron. This region is called a magnetic field.
4 A magnetic field is represented by lines with arrows on them. The arrows show the direction in which a N pole would move. Lines drawn close together represent a strong field.

# 20 ELECTROSTATICS

If a rubber balloon is rubbed with a duster it will attract dust particles. Perspex, cellulose acetate and the vinyl compounds used for gramophone records show the same attraction. This is because the materials have become charged with static electricity. Charging by friction is sometimes associated with a crackling sound. This is often heard when dry hair is combed or a terylene or nylon shirt is removed from the body. The crackling is caused by small electric sparks which may be seen if the room is in darkness.

If a rod is charged by rubbing and then touched on a small pith ball, the ball becomes charged. If two pith balls are suspended close to each other on nylon threads and charged from the same rod, they repel. However, if the balls are charged from two rods made of different materials they may attract each other. This behaviour indicates the presence of two types of static charge, referred to as **positive** and **negative** charge. It is found that **like charges repel and unlike charges attract**.

In experiments on static electricity the standard method for obtaining positive charge is to rub glass or cellulose acetate with silk. Negative charge is obtained on an ebonite rod by rubbing it with fur.

Glass, cellulose acetate, ebonite and polythene are examples of materials which are electrical insulators. That is, the charges produced on their surfaces do not move along or through the material, but remain at the spot where they are produced. If a piece of one of these materials is placed in an electric circuit, no current flows as the charge cannot pass through the piece of material.

Metals are examples of electrical conductors. Conductors can be charged by rubbing, for example by flicking fur across the cap of a gold-leaf electroscope. However, if a metal rod is held in the hand and rubbed, the charge formed flows through the metal and is lost to earth via the hand. If a piece of metal forms part of an electric circuit, the charge will pass through it and a current flows round the circuit. Some other materials conduct an electric charge but much less readily than metals. Only a very small current will pass round electric circuits containing such materials.

## 20.1 The gold-leaf electroscope

Figure 20.1 shows a common type of electroscope, used for the detection and testing of small quantities of an electric charge. It consists of a metal rod surmounted by a metal cap. A metal plate with a thin gold leaf attached is fixed to the lower end of the rod. The whole apparatus is protected from draughts by housing it in an earthed metal case with glass windows.

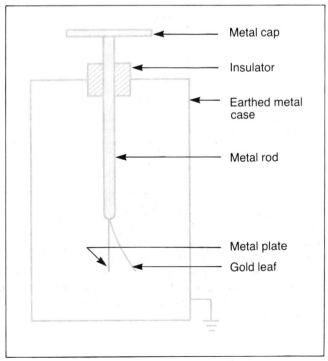

**Fig 20.1** Gold-leaf electroscope

If a rod of some suitable material is charged by friction and then brought near to the cap the leaf rises. If, for example, the rod is positively charged it will attract negative charges — from the atoms of the rod, plate and leaf — to the cap, leaving excess positive charges on the plate and leaf. The leaf and plate thus have the same charge, repulsion occurs and the leaf rises. This process is called **induction**, as the rod has not been brought into contact with any part of the gold-leaf electroscope, and no charge can therefore have been transferred from the rod to the electroscope. When the rod is removed from the cap, the leaf falls.

If the charged rod is scraped across or rolled over the surface of the cap, the leaf rises and remains up when the rod is removed. It may be necessary to repeat this procedure several times before this process of charging by contact is successful. Some of the charge on the surface of the rod has been transferred to the electroscope. If the cap of the charged electroscope is now touched with a finger, the charge flows to earth through the experimenter's body and the leaf falls. This is called **earthing**.

Figure 20.2(a) illustrates the method of charging by contact and (b) discharging by earthing.

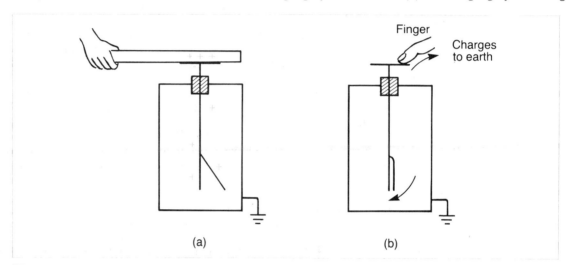

**Fig 20.2**

Successful charging of an electroscope by contact is difficult as the rod is an insulator and the charges do not flow over its surface. They have to be scraped off different areas by the cap of the electroscope. A more reliable way is to use the method of induction which is illustrated in Fig. 20.3.

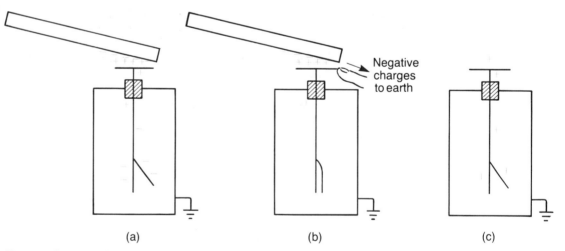

**Fig 20.3** Charging by induction

If an ebonite rod which has been charged by friction is brought close to the electroscope the leaf rises as shown in Fig. 20.3(a). With the rod still in place, the electroscope is earthed by touching any part of it which is conducting and the leaf falls (Fig. 20.3(b)). This happens because the negative charges previously on the plate and leaf are repelled to earth by the negative charges on the rod. However, the positive charges on the cap are still held in place by the attraction of the negative charges on the rod. The earth connection is removed. The rod is now removed and the positive charges distribute themselves throughout the electroscope (Fig. 20.3(c)); as a result the leaf rises.

## 20.2 Lines of force

In the same way that the force which a magnet exerts on a nearby compass needle can be represented by a magnetic field, so the force which a charged body exerts on a nearby charge can be represented by an electric field. An **electric line of force** is a line drawn in an electric field such that its direction at any point gives the direction of the electric force on a positive charge placed at that point. Thus lines of electric force begin on a positive charge and end on a negative charge of the same size. Figure 20.4 shows the lines of electric force in the vicinity of different arrangements of electric charges.

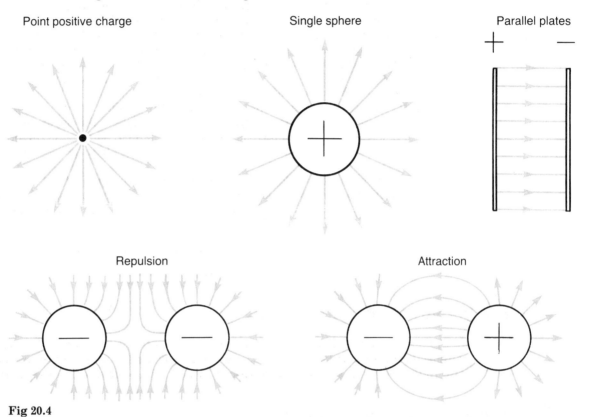

**Fig 20.4**

## 20.3 Summary

1 Like electrical charges repel, unlike charges attract.
2 An ebonite rod rubbed with fur, or a rubbed polythene strip, acquires a negative charge; a glass rod rubbed with silk or a rubbed cellulose acetate strip acquires a positive charge.
3 Metals, the human body and the earth are examples of conductors; glass, ebonite and plastics are normally insulators.
4 A conductor can be charged by contact or induction.
5 An induced charge can be obtained on an insulated conductor by: (i) bringing a charge near; (ii) touching the conductor; (iii) removing the finger; and (iv) taking the charge away.
6 Electric fields exist all round conductors which carry electric charges.

# 21 CURRENT ELECTRICITY

All atoms possess small negatively charged particles called electrons. In the case of metals some of these electrons are very weakly attached to their atoms and can easily be detached and made to flow through the metal. Metals are therefore good conductors of electricity.

In a battery one plate is at a positive potential and the other plate is at a negative potential, this potential difference being a property of the chemicals of which the battery is made. If the battery terminals are joined by a length of wire, the potential difference which exists between the ends of the wire will result in the weakly bound electrons already mentioned flowing through the wire. This constitutes an electric current.

Before the nature of an electric current was fully understood, the direction chosen to indicate current was, unfortunately, from the positive plate to the negative, that is, opposite to the flow of electrons. This is known as the *conventional current* and is the one normally used.

When the rate of flow of a charge past a point is $6 \times 10^{18}$ electrons per second, the current is 1 **ampere (A)**.

As an electric current is a flow of an electric charge, the quantity of electric charge which passes any point in a circuit will depend on the strength of the current and the time for which it flows. The unit of electric charge is the **coulomb. A coulomb is the quantity of electric charge conveyed in one second by a steady current of one ampere.**

Thus
$$\text{charge } (Q) = \text{current } (I) \times \text{time } (t)$$
$$Q = It$$

## 21.1 Potential difference

In order to achieve current flow in a circuit a potential difference $V$ must exist. The unit of potential difference is the **volt (V). Two points are at a potential difference of one volt if one joule of work is done per coulomb of electricity passing between the points.**

## 21.2 Electromotive force

The maximum potential difference that a cell is capable of producing in a circuit is called its **electromotive force (e.m.f.)**. It is measured in volts and may be regarded as the sum total of the potential differences which it can produce across all the components of a circuit in which it is connected, including the potential difference required to drive the current through itself. The e.m.f. is a property of the materials in the cell and is not affected by the nature of the circuit into which the cell is connected. On the other hand the potential difference which the cell produces in the external circuit connected to it depends on the current flowing through the circuit.

**The e.m.f. of a cell in volts is defined as the total work done in joules per coulomb of electricity conveyed in a circuit of which the cell is a part.**

## 21.3 Resistance

As the potential difference between the ends of a conductor is increased the current passing through it increases. **If the temperature of the conductor does not alter, the current which flows is proportional to the potential difference applied (Ohm's law).** Figure 21.1 shows this effect.

The gradient of this graph has a constant value, obtained by dividing the potential difference at any point by the current. The value of this constant gradient is known as the **resistance $R$** of the conductor. The unit of resistance is the **ohm ($\Omega$)**.

$$\frac{V}{I} = \text{constant } (R) \quad \text{(Ohm's law)}$$

hence $\quad V = IR$

A good conductor is one with a low resistance, a poor one has a high resistance.

In some conductors the current is not proportional to the potential difference between its ends. In a light bulb $V/I$ is found to increase with temperature, that is $R$ is not constant but

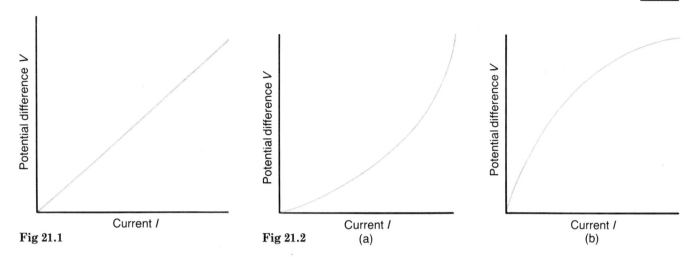

Fig 21.1

Fig 21.2 (a)

(b)

increases as the temperature of the filament of the bulb increases. Figure 21.2(a) shows the relationship between potential difference and current in such cases.

Figure 21.2(b) shows results that might be obtained from a thermistor for which the resistance decreases with increasing temperature.

## 21.4 Ammeters and voltmeters

An **ammeter** is used to measure the electric current through a conductor and must be in series with the conductor. It has a resistance; however, this should be as small as possible so that it only reduces the current to be measured by a very small amount.

A **voltmeter** measures potential difference between two points in a circuit. It must be in parallel with the part of the circuit concerned. In order that it should take as little current as possible out of the main circuit it must have as large a resistance as possible.

Figure 21.3 shows an ammeter Ⓐ and a voltmeter Ⓥ correctly connected to determine the value of the resistance $R$.

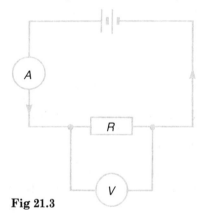

Fig 21.3

## 21.5 Resistors in series

A number of **resistors** $R_1$, $R_2$, $R_3$, are said to be connected **in series** if they are connected end to end as shown in Fig. 21.4. The current $I$ must be the same throughout the circuit as it has only one route to follow.

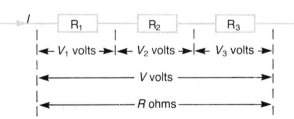

Fig 21.4 Resistors in series

If $R$ is the combined resistance of $R_1$, $R_2$, and $R_3$ and $V$ is the total potential difference across them, then
$$V = IR$$

but $V$ is the sum of the individual potential differences across $R_1$, $R_2$, and $R_3$. Thus
$$V = V_1 + V_2 + V_3$$
and
$$V = IR_1 + IR_2 + IR_3$$
therefore
$$IR = IR_1 + IR_2 + IR_3$$
and hence
$$\boldsymbol{R = R_1 + R_2 + R_3}$$

The same argument may be applied to any number of resistors in series.

In Fig. 21.3 the electromotive force $E$ of the battery must not only provide the potential difference to drive the current through the resistors, but also through itself $V_b$. The battery is said to have its own resistance $r$, known as its **internal resistance**. The total resistance of the circuit must include $r$, and is equal to the sum of $r$, $R_1$, $R_2$ and $R_3$ (Fig. 21.4). Now

$$V = V_1 + V_2 + V_3$$
and
$$E = V_1 + V_2 + V_3 + V_b$$
hence
$$E - V = V_b$$

$V_b$ is usually referred to as the *lost volts* in the circuit.

## 21.6 Resistors in parallel

Resistors are said to be **in parallel** when they are placed side by side in a circuit (Fig. 21.5).

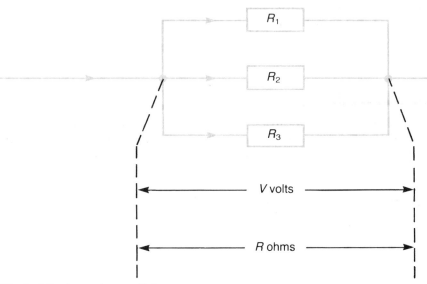

**Fig 21.5** Resistors in parallel

The conductance of each branch of the circuit is $I/V$, that is the quantity of current which passes for each volt of potential difference applied. But $I/V$ is equal to $1/R$; thus the conductance of each branch of a circuit is the reciprocal of its resistance. The total conductance of the three resistors in parallel is clearly the sum of the conductance of each individual resistor, as each one provides the current in the circuit with an alternative route, and thus makes its passage easier.

$$\text{Thus} \quad \frac{1}{R} = \frac{1}{R_1} + \frac{1}{R_2} + \frac{1}{R_3}$$

The same argument may be applied to any number of resistors in parallel. As an example consider a 2-ohm resistor and a 4-ohm resistor in parallel with each other.

$$\frac{1}{R} = \frac{1}{2} + \frac{1}{4} = \frac{3}{4}$$

$$\text{Thus} \quad R = \tfrac{4}{3} = 1\tfrac{1}{3}\,\Omega$$

The combined resistance of any number of resistors in parallel is always less than the value of any one of them. This is clear when one realizes that placing one resistor in parallel with another provides an alternative route for the current and thus eases its passage round the circuit.

## 21.7 Energy

When a potential difference is applied to the ends of a conductor some of the electrons inside it are set in motion by electric forces. Work is therefore done and the electrons acquire energy. The moving electrons form an electric current, and the energy of this current appears in various forms according to the type of circuit of which the conductor forms a part. For example in an electric fire the energy of the current is largely made available as heat, in an electric light as heat and light, and in an electric motor as mechanical energy of rotation. Energy in these various forms is produced at the expense of the source of electricity.

From the definition of the volt it can be seen that if a potential difference of one volt is applied to the ends of a conductor and one coulomb of electricity passes through it, then the work done, or energy transformed, is 1 joule. Hence if the potential difference applied is $V$ volts and the quantity of electricity that passes is $Q$ coulombs then the work done is $QV$ joules.

However   Charge $(Q)$ = current $(I)$ × time $(t)$

or   $Q = It$

therefore   work done = $QV = VIt$ joules

Two other expressions for the work done may be obtained by using Ohm's law. As

$$V = IR$$

$$\text{work done} = VIt = I^2 Rt = \frac{V^2 t}{R}$$

## 21.8 Power

The definition of power, given in Unit 3.11, is the rate of doing work or transferring energy.

$$\text{Power} = \frac{\text{work done}}{\text{time taken}}$$

If the work done is measured in joules, the unit of power is the **watt (W)**. Using the equations for energy given in the last unit, power may be obtained by dividing by time.

$$\text{Power} = VI = I^2 R = \frac{V^2}{R} \text{ watts}$$

The first expression in words is:

$$\text{Power} = \text{potential difference} \times \text{current}$$

*Example:*
Find (a) the current taken by, and (b) the resistance of, the filament of a lamp rated at 240 V, 40 W.

$$\text{Power} = \text{potential difference} \times \text{current}$$

therefore   $\text{current} = \dfrac{\text{power}}{\text{potential difference}}$

$$= \frac{40}{240} = \frac{1}{6} = 0.16 \text{ A}$$

but   potential difference = current × resistance

thus   $\text{resistance} = \dfrac{\text{potential difference}}{\text{current}}$

$$= \frac{240}{1/6} = 1440 \ \Omega$$

## 21.9 Cost

If an electricity meter is inspected it will be found to have the abbreviation 1 kWh on it. This stands for kilowatt-hour, the commercial unit of electrical energy. It is the energy supplied in one hour by a rate of working of 1000 W.

$$1 \text{ kilowatt-hour} = 1000 \text{ watt-hours}$$

$$= 1000 \text{ joules/second for 1 hour}$$

$$= 1000 \times 60 \times 60 \text{ joules}$$

$$= 3\,600\,000 \text{ joules or } 3.6 \text{ MJ}$$

*Example:*
Find the cost of running a 2 kW fire for 3 hours if the cost of electrical energy is 6.0 pence per kilowatt-hour (unit).

$$\text{Total energy consumed} = \text{power} \times \text{time}$$

$$= 2 \times 3 \text{ kWh}$$

$$= 6 \text{ kWh}$$

$$\text{Cost} = 6 \times 6.0 = 36 \text{ pence}$$

## 21.10 House electrical installation

The cable bringing the mains electricity supply into a house contains two wires, one of which is 'live', and the other 'neutral'. The neutral wire is earthed at the local transformer substation, so it is at earth potential. At some convenient place inside the house the mains cable enters a sealed box, where the live wire is connected to the Electricity Board's fuse. On the far side of this box the power cable enters the meter and from there it goes to the main fuse box. The fuse box contains a separate fuse for each of the lighting circuits, ring circuit and cooker circuit (Fig. 21.6).

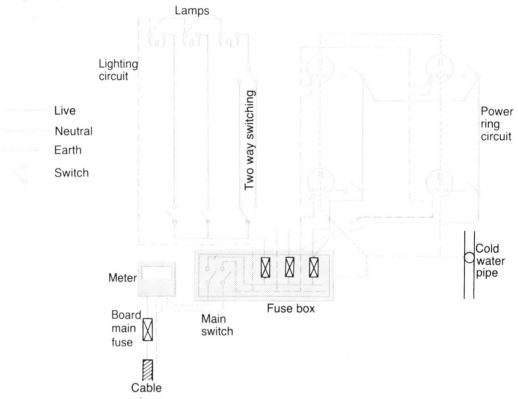

**Fig 21.6** House electrical installations

In modern installations the power sockets are tapped off a ring circuit. This cable passes through the various rooms in the house and has both its ends connected to the mains supply. Thus there are two paths by which the current may get to a particular socket, which effectively doubles the capacity of the cable.

It can be seen from Fig. 21.6 that all light and power switches and fuses are placed in the live side of the supply. If they were in the neutral side all light and power sockets would remain live when the switches were in the off position.

## 21.11 Fuses

For safety reasons all domestic electrical appliances are fused. That is, in at least one place on each circuit, a fuse is fitted, which will 'blow' if a fault develops in the circuit and too great a current passes through it. This arrangement protects the wiring against the possibility of overheating and setting fire to the house.

The main house fuses for the different circuits generally consist of short lengths of tinned copper wire fitted into porcelain carriers. These fuses are usually situated under the stairs or in a cupboard which also contains the electricity meter and Electricity Board fuse.

In addition, items which are separately plugged into power sockets — for example, fires — contain a fuse in the plug. This type of fuse consists of a small glass cartridge with a thin wire through its centre. Such a fuse is rated at 2, 5, 10 or 13 A to suit the appliance to which it is connected. For example, the value of the fuse which should be used in the plug of a 240 V, 2 kW fire is calculated as follows.

$$\text{Power} = \text{voltage} \times \text{current}$$
thus
$$2000 = 240 \times \text{current}$$
and the
$$\text{current} = \frac{2000}{240} \text{ A} = 8.33 \text{ A}$$

A 10 A fuse would thus be suitable.

If a fault develops in the appliance, this fuse will normally 'blow' rather than the larger one in the main fuse box. This avoids other appliances being put out of action at the same time.

The thickness of wire used in any fuse is such that it will overheat and melt if the current passing through it exceeds its specified rating by much. Once the wire has melted the circuit is broken.

When a fuse has 'blown' it is essential that the cause of the fuse blowing is found before it is repaired or replaced and the appliance used again. Sometimes a fuse 'blows' because the fuse wire is very old and has become weakened by oxidation, or it can blow due to a fault in the circuit, such as a short circuit in the flex where the insulation has worn and frayed. Whatever the fault it must be found and put right before a new fuse is fitted.

## 21.12 Earthing

Besides the live (brown) and neutral (blue) wires, all power circuits are provided with a third wire (green and yellow stripes) which has been earthed by a good electrical joint to the cold water supply. When an appliance is connected to the circuit the earth wire provides a low resistance route between the casing of the appliance and the earth. This is a safeguard to prevent anyone receiving a shock by touching the casing should this become 'live'. Such a danger would arise if the insulation on the live flex had become worn and allowed the live wire to come into contact with the casing of the appliance. If this happened in a properly earthed appliance, a large current would instantaneously flow to earth and the fuse would blow, thereby cutting off the supply.

Loose connections in plugs are another potential source of danger to the person. Proper earthing again removes the danger.

## 21.13 Summary

1. **Charge ($Q$) = current ($I$) × time ($t$).**
2. In order to achieve current flow a potential difference must exist. The unit of potential difference is the volt (V). Two points are at a potential difference of one volt if one joule (J) of work is done per coulomb of electricity passing between the points.
3. The resistance of a conductor is defined by the equation

$$\textbf{Resistance } (\boldsymbol{R}) = \frac{\textbf{potential difference } (\boldsymbol{V})}{\textbf{current } (\boldsymbol{I})} \quad \text{(Ohm's law)}$$

Current is measured in amperes (A) and resistance in ohms ($\Omega$).

4. A good conductor has a low resistance and a poor conductor has a high resistance.
5. In a simple series circuit the current has the same value at all points round the circuit.
6. The total resistance $\boldsymbol{R}$ of a number of resistors $\boldsymbol{R_1}, \boldsymbol{R_2}, \boldsymbol{R_3}$, in series is given by

$$\boldsymbol{R} = \boldsymbol{R_1} + \boldsymbol{R_2} + \boldsymbol{R_3} \text{ etc.}$$

7. In a parallel circuit the total current entering a junction must equal the total current leaving it.
8. The total resistance $\boldsymbol{R}$ of a number of resistors $\boldsymbol{R_1}, \boldsymbol{R_2}, \boldsymbol{R_3}$ all in parallel with each other is given by

$$\frac{1}{\boldsymbol{R}} = \frac{1}{\boldsymbol{R_1}} + \frac{1}{\boldsymbol{R_2}} + \frac{1}{\boldsymbol{R_3}} \text{ etc.}$$

9. The work done or energy transformed by the flow of an electric current is given by

$$\textbf{work done} = VIt = I^2Rt = \frac{V^2 t}{R}$$

10. Power is the rate of doing work or transforming energy

$$\textbf{Power} = \frac{\textbf{work done}}{\textbf{time taken}} = VI = I^2R = \frac{V^2}{R}$$

11. The fuse is the weakest point in a circuit. The fuse wire melts and breaks the circuit if more current flows than the circuit is designed to take. The fuse protects the appliance and the circuit from damage.
12. The earth wire provides a low resistance route between the casing of an appliance and earth. It prevents anyone receiving a shock by touching the casing should the casing become 'live'.

# 22 ELECTROMAGNETISM

Whenever an electric current flows in a conductor a magnetic effect is present in the region of the conductor. This can most easily be demonstrated by passing a current down a vertical conductor, such as a retort stand (Fig. 22.1).

At some convenient height the conductor passes through a hole in a horizontal platform. A number of small compasses are placed on the platform.
When the current is switched on the compass needles swing from pointing to magnetic north and re-align themselves in a circular path around the wire.
The influence of the current in the region of the wire is called a **magnetic field**.

This result can be remembered using the following rule. Imagine the wire to be grasped in the right hand with the thumb pointing along the wire in the direction of the current. The direction of the fingers will give the direction of the lines of force. This rule is purely an aid to memory; it does not explain how the lines occur.

A knowledge of the magnetic field around a straight conductor can be used to predict the pattern round other simple geometrical arrangements of current carrying wires — for example, a flat coil and a solenoid.

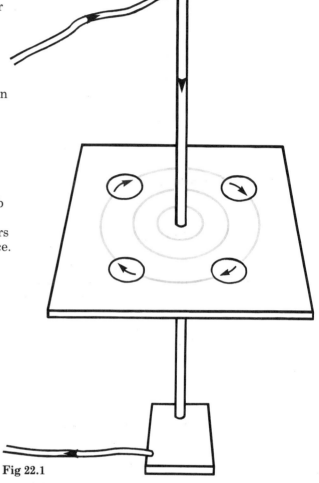

Fig 22.1

The pattern of lines of force when a current is passed through a flat circular coil is shown in Fig. 22.2.

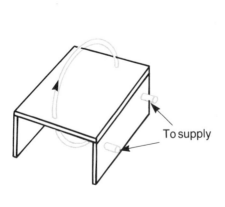
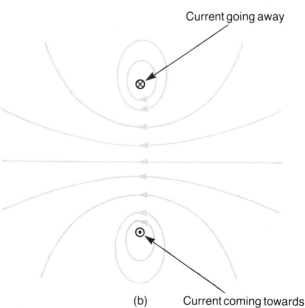

Fig 22.2    (a)    (b)

Figure 22.3 illustrates the magnetic field due to a current in a solenoid.

If one considers the circular magnetic field round each short length of wire in the flat circular coil, it can be seen that the field adds up through the centre of the coil. This leads to a strong field through the coil but a weak one outside it (Fig. 22.2(b)). A solenoid may be considered as a series of flat circular coils, each a little spaced from one another on a common axis. Each turn of insulated wire gives a magnetic field similar to that of a flat circular coil. The fields between neighbouring turns oppose one another and cancel, but the fields along the common axis reinforce, producing the pattern shown in Fig. 22.3(b).

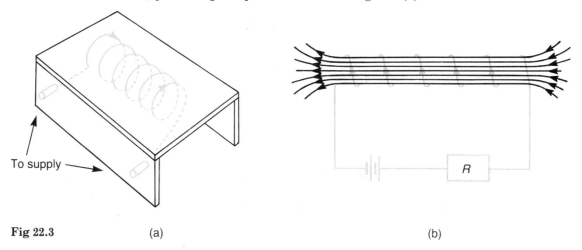

**Fig 22.3**  (a)  (b)

The magnetic fields resulting from a current passing through flat coils and solenoids of square cross-sectional area are similar to those shown for circular areas.

## 22.1 The electromagnet

The use of a solenoid for making magnets has already been described in Unit 19.1. When a piece of steel is placed inside a solenoid and the current switched on, the steel becomes a magnet and remains one when the current is switched off. If a piece of soft-iron is used instead, the iron acts as a magnet only while the current is switched on. Figure 22.4(a) shows such an electromagnet. Sometimes the cores of electromagnets are U-shaped (Fig. 22.4(b)). This arrangement has the advantage that the attraction of both ends can be used simultaneously.

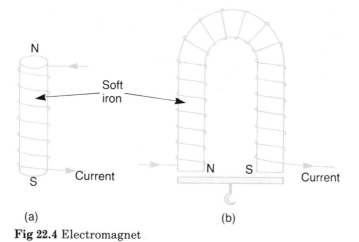

(a) (b)

**Fig 22.4** Electromagnet

Electromagnets are very strong. A small one, made in a laboratory from a U-shaped core and a few turns of wire each carrying a current of a few amps, is capable of lifting several kilograms. Those used in industry can lift several tonnes, for example car and lorry bodies. Yet when the current is switched off, the magnetism ceases and the load is released.

## 22.2 The electric bell

The electric bell consists of two solenoids wound in opposite directions on two soft-iron cores joined by a soft-iron yoke (Fig. 22.5). One end of the windings is connected to the battery and the other to a metal bracket which supports a spring-mounted soft-iron armature (moving part). The armature carries a light spring to which is soldered a small platinum disc as contact.

The disc presses against the end of a platinum-tipped contact screw from which a wire goes to the push switch. When the switch is pressed a current flows through the circuit and the cores become magnetized. The resultant attraction of the armature separates the contacts and breaks the circuit.

The cores become demagnetized and the armature is returned to its original position by the spring. Contact is re-made and the action repeated. As a result the armature vibrates and a hammer attached to it strikes the gong.

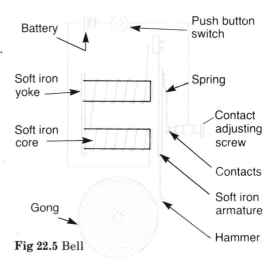

**Fig 22.5** Bell

## 22.3 The force on charges moving in a magnetic field

Figure 22.6 shows a current-carrying wire placed in a magnetic field. As soon as the current is switched on the length of wire $AB$ moves horizontally on the other two pieces. The interaction of the magnetic field due to the current flow and that due to the magnets, results in a horizontal force. The effect may be understood by reference to Fig. 22.7.

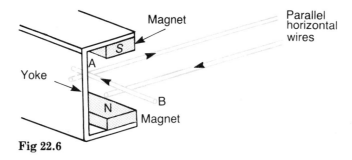

**Fig 22.6**

The two fields shown separately in Figs. 22.7(a) and (b) reinforce to the left of the wire whereas they tend to cancel to the right (Fig. 22.7(c)). A strong magnetic field results on the left with the lines in tension. The tension results in a force from left to right. If either magnetic field is reversed the force reverses.

It should be noted that the current, the magnetic field due to the magnets and the force produced are all three mutually at right angles. If the current is parallel to the field due to the magnets there is no force.

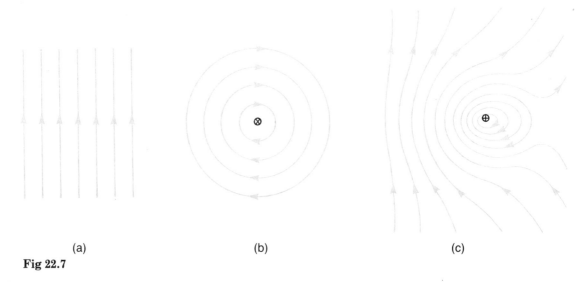

(a)                    (b)                    (c)

**Fig 22.7**

If a stream of electric charges passes through a gas, as in a fine beam tube (Fig. 22.8), rather than a wire, the same force results. However, whereas in a wire the charges (electrons) are inhibited from moving by its stiffness, this is not so in a gas, where they are free to change their

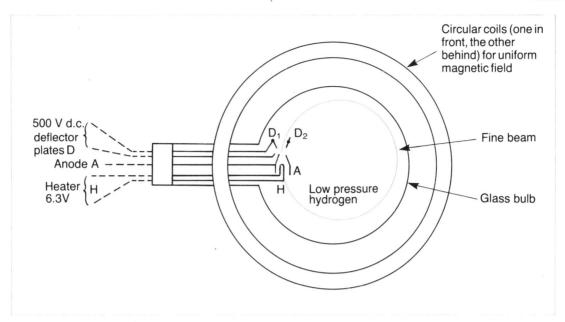

**Fig 22.8** Fine beam tube

direction of motion as soon as the force acts. The force always remains at right angles to their direction of motion, as well as to the magnetic field, and the charges move in the arc of a circle (see Unit 3.4).

## 22.4 The d.c. electric motor

Figure 22.9 shows the construction of a simple direct current (d.c.) electric motor. It consists of a rectangular coil of wire mounted on a spindle so that it is free to rotate between the pole pieces of a permanent magnet. The two ends of the coil are soldered to the two halves of a copper split ring or **commutator**, which is a device for reversing the direction of current flow in the coil every half revolution. The two brushes press lightly against the commutator. When a current is passed through the coil it rotates. Figure 22.10(a) shows a cross-section through the coil together with the resultant magnetic field, which is strong above the right side and also below the left side. The coil therefore rotates clockwise.

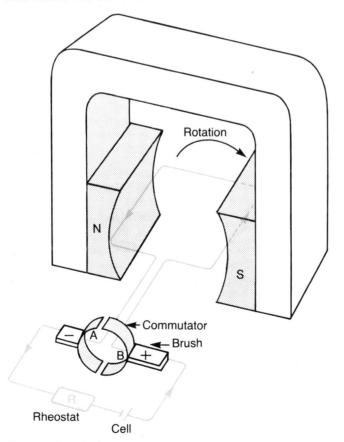

**Fig 22.9** Simple d.c. motor

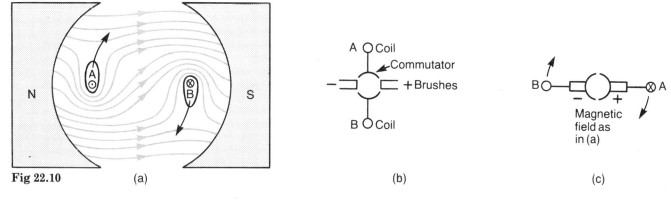

Fig 22.10

When the coil is vertical the brushes touch the space between the two halves of the commutator. There is no torque at this position (Fig. 22.10(b)); however, the coil's momentum carries it past the vertical, and when this has occurred the commutator halves automatically change contact from one brush to the other. This reverses the current through the coil (Fig. 22.10(c)) and thus it rotates through the next half turn. The reversal of the direction of current flow by the commutator each half turn ensures the continued rotation of the coil.

The simple motor described is not very efficient or powerful, as the torque changes from a maximum when the coil is horizontal to zero when it is vertical. The motor can be improved by winding a number of coils, each of many turns, at different angles round a soft-iron armature. The greater number of turns in a coil gives a greater torque; having a number of coils ranged at angles round the armature means that when one is vertical another is horizontal thus resulting in an even torque. The iron armature becomes magnetized and increases the magnetic field through the coils, resulting in a greater torque. It is also usual for the magnetic field to be provided by electromagnets rather than permanent magnets.

## 22.5 The moving coil meter (galvanometer)

The moving coil meter relies on the same principle for its operation as the simple electric motor just described. The coil is normally mounted vertically on a phosphor bronze suspension strip or spring rather than horizontally on a spindle. A loosely coiled spring below the coil controls how far it turns; the greater the current through the coil the greater the angle through which it turns before the torque exerted by the spring equals the torque due to the magnetic field. The reading of the meter is shown by attaching a pointer to the coil and allowing it to move over a scale.

## 22.6 Summary

1. Whenever an electric current flows in a conductor a magnetic field is present in the region of the conductor.
2. The magnetic field round a coil carrying direct current is similar in shape to that round a bar magnet.
3. When a steel bar is placed in a coil carrying direct current the bar becomes permanently magnetized.
4. When a bar of soft iron is placed in a coil carrying direct current the soft iron becomes a strong magnet only while the current flows. This is an electromagnet.
5. Charges moving at 90° to a magnetic field experience a force at 90° to both their direction of motion and the direction of the magnetic field.
6. The behaviour described in paragraph 5 is the basis of an electric motor.

# 23 ELECTROMAGNETIC INDUCTION

The magnetic field in a region is often referred to as **magnetic flux**.

## 23.1 Laws

**1 Faraday's. Whenever there is a change in the magnetic flux linked with a circuit an electromotive force is induced, the strength of which is proportional to the rate of change of the flux through the circuit.**

**2 Lenz's. The direction of the induced current is always such as to oppose the change producing it.**

The truth of these laws is best illustrated by considering a simple experiment. First a centre-zero meter is connected in series with a cell and a suitable high resistance, and the direction of movement of the pointer noted when a small current passes in a known direction. The meter is now connected to the ends of a straight wire placed at right angles to the lines of magnetic field between two opposite magnetic poles (Fig. 23.1).

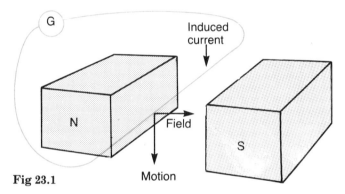

**Fig 23.1**

If the wire is moved downwards the meter indicates that an induced current flows in the direction shown. If the wire is moved up the current flows in the opposite direction. The same results are obtained if the wire is held still and the magnets moved up or down, respectively. It is the relative motion between the wire and magnetic field which leads to the induced voltage and hence current. The quicker the relative motion takes place the greater the deflection (Faraday's law).

While the induced current flows there is a magnetic field around the wire. Consideration of the effect of this field and that of the magnets shows that it results in an upward force in Fig. 23.1. That is, the induced current leads to a force acting in the opposite direction to the movement of the wire and thus opposing it (Lenz's law).

## 23.2 The simple d.c. dynamo

The simple direct current electric motor described in Unit 22.4 may be used as a simple d.c. dynamo. The motor is connected in series with a resistance and a moving coil meter instead of a voltage source. When the coil is rotated the meter is seen to deflect in one direction, although the deflection is not a steady one. The commutator ensures that, although the current in the coil itself reverses during the second half of a rotation, the same brush always remains positive and the other negative.

The simple dynamo just described is not very efficient. A practical dynamo has a number of coils wound in slots cut in the armature. Each coil has its own pair of segments in a multi-segment commutator. This arrangement ensures that the e.m.f. obtained is fairly steady, as it is only the horizontal coils which are connected to the brushes at any given moment. The iron armature is built in layers, each one insulated from its neighbours. Although e.m.f.'s are induced in these layers of iron as they rotate in the magnetic field, very little current flows in the armature as a result, due to this insulation.

## 23.3 The simple a.c. dynamo

This differs from the simple d.c. version, described in the last unit, only in its connections (Fig. 23.2). The ends of the coil are connected to two slip rings mounted on the coil spindle. One side of the coil is thus always connected to the same brush.

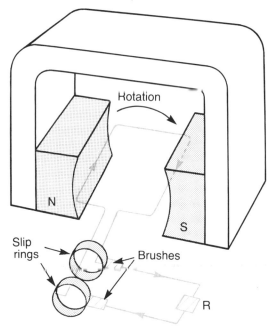

**Fig 23.2** Simple a.c. dynamo

The outputs of both a simple a.c. and a simple d.c. dynamo are shown in Figs 23.3(a) and (b), respectively. In each case the coil starts from the vertical.

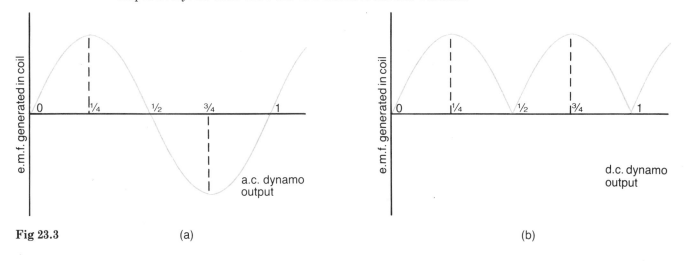

**Fig 23.3**         (a)         (b)

## 23.4 Alternating current

Figure 23.3(a) shows how the voltage from an a.c. dynamo changes with time. One revolution of the dynamo produced one complete wave or one cycle as shown in the diagram. The number of complete waves produced every second is known as the frequency of the alternating current. The unit of frequency is the **Hertz (Hz)**. In Great Britain the frequency of the mains supply is 50 Hz.

The maximum value of an alternating voltage during its cycle is known as its **peak value**. However, for much of the time, the magnitude of the voltage is considerably less than this, and an alternating voltage of a particular peak value will supply energy at a slower rate than a steady voltage of the same value. It is thus usual to express the root-mean-square (r.m.s.) value of the voltage which is related to the peak voltage by the equation

$$E_{r.m.s.} = \frac{E_p}{\sqrt{2}}$$

The **root-mean-square voltage** is the steady voltage that would produce the same rate of energy supply as the alternating voltage of peak value $E_p$. In Great Britain the 240 V a.c. mains voltage means that 240 V is the r.m.s. voltage of the mains. The peak voltage $E_p$ is 240 $\times \sqrt{2} = 340$ V. The 240 V a.c. mains thus produces the same power as would be obtained from a steady voltage of this value.

## 23.5 The transformer

A transformer is illustrated diagrammatically in Fig. 23.4. The two coils shown are wound on a laminated soft-iron core. When an alternating current passes through the primary an alternating magnetic field is set up in the core. Since the flux through the secondary coil is changing, this induces an e.m.f. in it which is also alternating. The size of this induced e.m.f. will depend on the e.m.f. applied to the primary and on the relative numbers of turns in the two coils:

$$\frac{\text{secondary e.m.f.}}{\text{primary e.m.f.}} = \frac{\text{number of turns in secondary}}{\text{number of turns in primary}}$$

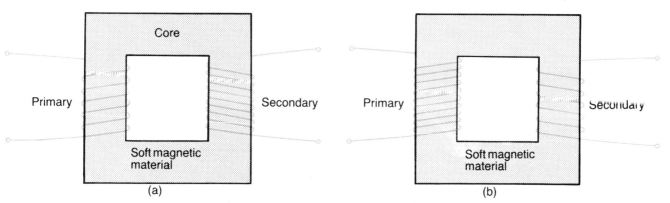

**Fig 23.4** Transformer

If the ratio of the number of turns is greater than one, the secondary e.m.f. will be greater than the primary and we have a *step-up* transformer. A ratio less than unity gives a *step-down* transformer.

A transformer is designed to minimize energy losses. The windings are composed of low-resistance copper coils to reduce the heating losses. The core is laminated, that is constructed of layers of magnetic material fixed together and insulated from each other by varnish or oxide coatings. This ensures that, although e.m.f.'s are induced in the core, as well as the secondary coil by the changing magnetic field, currents cannot flow between the laminated layers. Thus the total current in the transformer core is kept to as small a value as possible, as are the energy losses due to its heating effect.

Efficient core design also means that all the magnetic flux produced by the primary passes through the secondary. In practice this is best achieved by winding the primary and secondary on top of one another. In a good transformer energy losses are small and we may assume that:

$$\text{secondary power output} = \text{primary power input}$$

## 23.6 Power transmission

One of the main advantages of alternating current is that it can easily be changed from one voltage to another by a transformer, with little loss of energy. For this reason, electric power is generally conveyed by alternating current, as it can be transformed to a very high voltage and transmitted over large distances, with small power losses as explained below. This has two advantages. Firstly, electricity can be generated where water power, coal and oil are easily obtainable or at conveniently sited nuclear power stations, and conveyed to all parts of the country by high voltage overhead power lines. Secondly, power is easily made available wherever the peak demand occurs. For example, during the day power which has been generated in an area that has little industry may be used in a more industrialized area that has at this time a greater demand.

In great Britain electricity is generated at 11 000 V and then stepped up to as much as 400 000 V by transformers. It is subsequently stepped down in stages at substations in the neighbourhood where the energy is to be consumed. The reason electrical energy is transmitted at such high voltages can clearly be seen from the following calculation.

Find the power wasted as heat in the cables when 10 kW is transmitted through a cable of resistance 0.5 Ω at (a) 200 V; (b) 200 000 V.

(a) The current is given by the equation:

$$\text{current} = \frac{\text{power}}{\text{voltage}} = \frac{10\,000}{200}$$

$$= 50 \text{ A}$$

Therefore the power lost in the cable $= I^2 R$

$$= 50^2 \times 0.5 \text{ W}$$

$$= 1250 \text{ W}$$

(b) The current $= \dfrac{\text{power}}{\text{voltage}} = \dfrac{10\,000}{200\,000} = 0.05$ A

Therefore the power lost in the cable $= I^2 R$

$$= 0.05^2 \times 0.5 \text{ W}$$

$$= 0.00125 \text{ W}$$

At 200 V more than 10% of the energy transmitted is wasted in heating the cable. At 200 000 V this energy loss is reduced by a factor of a million and is negligible.

## 23.7 Summary

1. Whenever a conductor moves relative to a magnetic field an electromotive force (e.m.f.) is induced in the conductor. The strength of the e.m.f. is proportional to the strength of the magnetic field and the speed of the relative motion.
2. The effect outlined in paragraph 1 is the basis of a dynamo.
3. A simple dynamo (with no commutator) generates an alternating e.m.f. (voltage).
4. A transformer changes the value of an alternating e.m.f. either up or down.
5. For a transformer

$$\frac{\text{secondary e.m.f.}}{\text{primary e.m.f.}} = \frac{\text{number of turns in secondary}}{\text{number of turns in primary}}$$

6. Power is transmitted over long distances at high voltage and low current. This reduces energy losses in the transmission cables.

# 24 ELECTRON BEAMS

## 24.1 Thermionic emission

Metals contain many electrons which are loosely attached to their atoms. If a wire is heated to a high temperature the extra energy given to the electrons enables them to break away from the metal structure and exist outside as an electron cloud. This is called **thermionic emission**.

Strontium and barium are good electron emitters but are not suitable to be made into a thin wire. Tungsten, however, can be made into a very thin wire and can withstand high temperatures without melting. Thus tungsten is used as the base of the wire and its surface coated with barium or strontium which emit well at these temperatures. The wire is heated electrically, usually by using a 6.3 V supply which can be obtained from either a d.c. or an a.c. source.

## 24.2 The diode

The electrons released from a wire by thermionic emission form a cloud around it which prevents further emission. This cloud may be removed by placing a plate near to the wire and connecting a steady voltage between them, so that the plate (anode) is more positive than the wire (cathode). It is necessary to place this whole arrangement in an evacuated tube so that the passage of the electrons is not inhibited by the presence of air molecules. This arrangement is known as a **diode**.

The characteristic action of a diode may be investigated using the circuit shown in Fig. 24.1(a). The voltage applied to the diode is varied using the source $B$, and the resulting current is registered on the milliammeter. Figure 24.1(b) shows a typical set of results.

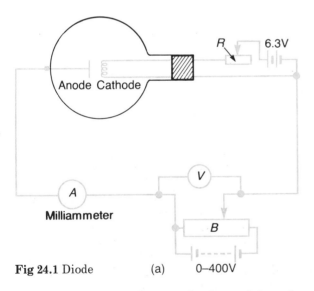
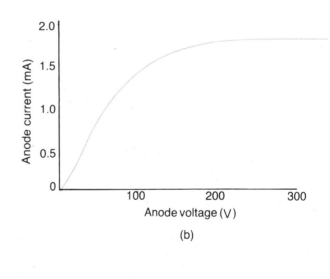

**Fig 24.1** Diode

The actual values of the voltage applied and the current obtained will depend on the precise diode used. However, the general shape of the graph will be similar in all cases. When no voltage is applied between the anode and cathode the electrons released by thermionic emission remain near the cathode. If a small positive potential difference is applied between the anode and cathode, some of the electrons in the cloud move across the empty space to the anode. Here they flow through the anode circuit back to the cathode, and a small current is registered.

The electrons crossing to the anode create a negative charge in the space between anode and cathode. This negative space charge repels some electrons back to the cathode. Thus not all the electrons emitted from the cathode reach the anode when the potential difference is small. However, as the potential difference increases, a larger proportion of the electrons emitted do reach the anode. At large voltages all the electrons emitted per second reach the anode and increasing the voltage further will not increase the current. The maximum current is known as the *saturation current*.

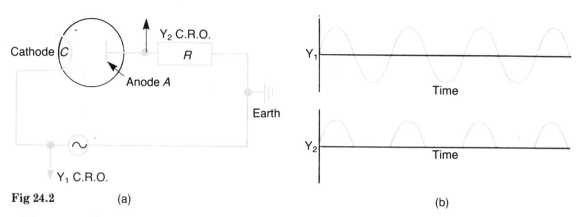

**Fig 24.2** (a) (b)

No current flows when the anode potential is negative with respect to the cathode. All the electrons emitted are repelled back to the cathode. The diode will thus only allow current to flow in one direction; for this reason it is known as a **valve**.

Suppose an alternating voltage is connected to a diode valve with a resistance $R$ of a few thousand ohms in series with it (Fig. 24.2(a)). The diode will only pass current during the positive half of each cycle. An oscilloscope connected across the resistance $R$ will show a voltage of the form illustrated in Fig. 24.2(b). The applied voltage has been rectified by the diode valve which has been used as a 'rectifier'.

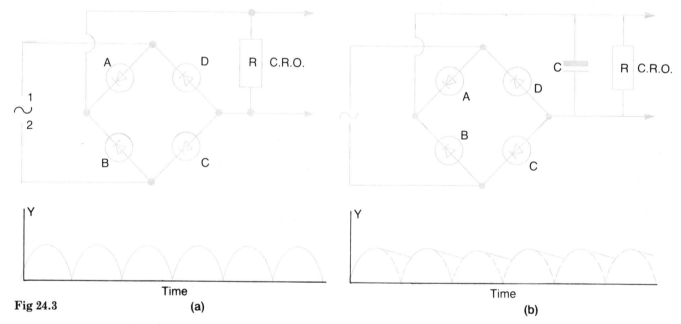

**Fig 24.3** (a) (b)

One diode only allows half of an alternating current to flow through a load $R$; the other half is cut out. A more even supply of power can be obtained using the circuit in Fig. 24.3(a). This makes the whole of the alternating current flow one way through $R$. When terminal 1 is positive, current flows through diode $A$, the resistor $R$ and diode $C$ and returns to terminal 2. When terminal 2 is positive, current flows through diode $B$, the resistor $R$ and diode $D$ before returning to terminal 1. The output waveform is shown under the circuit diagram.

A smoother output may be obtained by using the circuit shown in Fig. 24.3(b). This is the same as that shown in Fig. 24.3(a) with the addition of a smoothing capacitor $C$ in parallel with the resistor $R$. While the potential difference across $R$ and $C$ is rising to its peak value, the capacitor $C$ is charged up. Then while the output of the rectifier drops rapidly to zero, the capacitor supplies charge, causing the current through $R$ to fall more slowly. The larger the value of the capacitor, the smoother is the final output. The output waveform is shown under the circuit diagrams in Fig. 24.3(b).

## 24.3 Cathode rays (electron beams)

Since electrons are easily produced by thermionic emission, experiments on electrons can be conveniently carried out using this source.

In the Maltese cross tube illustrated in Fig. 24.4 the anode $A$ is cylindrical and is maintained at a positive potential of a few thousand volts relative to the cathode $C$. This part of the apparatus is called an *electron gun*. The electrons are accelerated in the space between the cathode and anode and pass through the cylindrical anode. The Maltese cross shown is at right angles to the beam of cathode rays and is connected to the anode. The cross is thus at the same

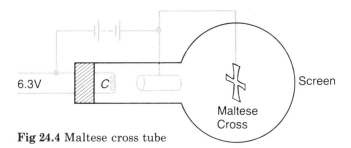

**Fig 24.4** Maltese cross tube

potential as the anode and the electrons move between the two at a constant speed (that is, they are not accelerated). When the electrons strike the screen some of their kinetic energy is converted into light, thus showing the position of the beam. A sharp shadow of the cross is seen on the screen, suggesting that the electrons emitted from the cathode travel in straight lines along the tube.

## 24.4 The effect of electric and magnetic fields

The deflection tube illustrated in Fig. 24.5 is similar to the Maltese cross tube, except that it contains two horizontal plates $P$ and $Q$ instead of the cross. The electrons emitted from the cathode are accelerated in the same way. They then pass between $P$ and $Q$ before striking the centre of the screen. However, if a large potential difference is connected between $P$ and $Q$, the electron beam will be deflected. If, for example, $P$ is positive relative to $Q$ the electrons will be attracted to it and the beam will be deflected towards it. The beam is deflected downwards if $Q$ is the more positive plate. The amount of deflection depends on the potential difference.

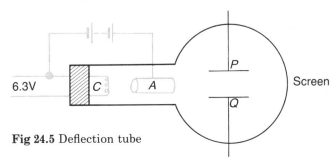

**Fig 24.5** Deflection tube

If two bar magnets are placed, one each side of the tube, with opposite poles facing, the beam will be deflected. This is the same effect as that responsible for the force on a current-carrying wire placed in a magnetic field and thus for the rotation of an electric motor. In place of the bar magnets, the magnetic field may be produced by using two vertical coils, one each side of the tube. If a current of between one and two amps is passed through these coils connected in series, a magnetic field results in the space between them. It has already been seen, in Unit 22.3, that a beam of charged particles passing through a magnetic field will experience a force at right angles to the field and their direction of travel. In the deflection tube the force acting on the electrons is initially either up or down, but, as the beam is deflected by this force, the force remains at right angles to the beam. The amount of deflection in a magnetic field depends on the strength of the field; in the case of the coils this depends on the current passing through them.

There is one further common type of deflection tube, known as the fine beam tube (Fig. 22.8). It contains a conical metal anode $A$, with a hole at the top over the cathode $C$. When a potential difference of a few thousand volts is connected between the two an electron beam is emitted vertically. The tube has a very small amount of hydrogen in it. The fast-moving electrons produce a fine beam of light as they ionize the hydrogen molecules. The beam shows the path of the electrons. The electrons may be deflected by connecting a potential difference between the two plates $D_1$ and $D_2$ just above the anode, or by passing an electric current through the two large coils placed one each side of the tube. In the latter case the beam of electrons may be deflected into a closed circle as the force is always at right angles to the direction of travel of the electrons.

## 24.5 Cathode ray oscilloscope

The cathode ray tube shown in Fig. 24.6 is very similar to the deflection tube illustrated in Fig. 24.5. It relies on exactly the same principle for its operation but has an additional pair of plates $X_1$ and $X_2$ so that the electron beam may be deflected horizontally as well as vertically.

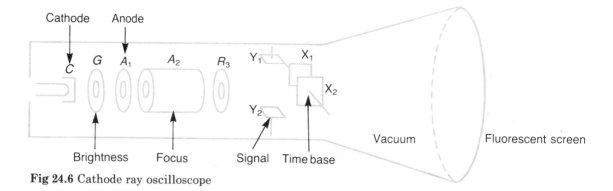

**Fig 24.6** Cathode ray oscilloscope

The cathode $C$ emits electrons which are accelerated to the anode $A$ by a high positive potential difference. In practice the anode usually consists of more than one plate or cylinder ($A_1$, $A_2$) so that it behaves like an electron lens and is able to focus the beam. The plate $G$ is slightly negative compared to the anode. The number of electrons reaching the screen is controlled by how negative $G$ is.

The cathode ray oscilloscope is excellent for use as a voltmeter. As no charge passes between the plates, when the beam is deflected, the tube draws no current from the component whose voltage it is recording; that is, it has an extremely high resistance.

If a d.c. voltage is to be measured it is connected between the plates $Y_1$ and $Y_2$. The spot on the screen will move a certain distance vertically which should be measured. If the oscilloscope has a calibrated scale this reading may immediately be converted to volts. If it is not calibrated a battery of known voltage should be connected to the oscilloscope and the deflection it causes recorded.

If it is required to measure an a.c. voltage, this should be connected to the plates $Y_1$ and $Y_2$. The length of the vertical line which results gives the value of the peak to peak voltage. However, if it is desired to see how the value of the voltage changes with time, the time base facility should be used. This provides a changing voltage connected to the $X$-plates, thus deflecting the beam horizontally. The spot thus repeatedly moves across the screen at a speed dictated by a switch controlling the time base frequency. Thus, as the spot moves up and down in response to the a.c. voltage connected to the $Y$-plates, it moves horizontally at a steady speed. A graph (Fig. 24.2(b) $Y_1$) is obtained with voltage as the $Y$-axis and time as the $X$-axis.

The cathode ray oscilloscope may be used to display any a.c. voltage; for example that across a resistor or that given by a microphone when sound falls on its diaphragm. It may also be used to examine the small voltage associated with the human heartbeat. The time base mechanism makes the oscilloscope suitable for use as a timing device. A pulse can be displayed on the screen at the instant it is emitted by a radar system and again when it is received back after reflection by an object in the earth's atmosphere or in the space beyond. The distance between the two pulses on the screen gives the time taken for the return journey of the pulse. As the velocity of radio waves is known to be $3 \times 10^8$ m/s, the distance between the emitter and the reflecting object can be calculated.

The cathode ray tube is the basis of a television set. The time base causes the spot to cross the screen many times in quick succession, each time a little below the last, so that the whole screen is covered. In Great Britain and many other countries 625 lines are drawn on the screen in 1/25 second in this way. As the spot moves across the screen its intensity is varied by the incoming signal, thus causing a picture to be 'painted' on the screen. The whole process is then repeated. Thus 25 slightly different pictures appear on the screen every second, giving the impression of continuous movement.

## 24.6 Semiconductor materials

Materials which allow electrons to flow through them are called conductors. Metals are the best conductors. The outermost electrons in each atom are so loosely attached that they are able to move freely between atoms. They are called 'free electrons'.

Materials which do not conduct charge are called insulators. All their electrons are tightly held to atoms and are not normally free to move. Glass, rubber, polythene and plastics are good examples of insulators.

A few materials are neither good conductors nor good insulators, but have conducting properties between the two groups. These materials, which are known as semiconductors, contain a small number of free and mobile charges. The best known semiconductor material is silicon although there are others, such as germanium, lead sulphide and gallium arsenide.

In a pure semiconductor such as silicon, all the outer electrons of the atoms form bonds with neighbouring atoms and there are virtually no 'free electrons' left over for conduction. However, as the temperature is increased, the thermal energies of the vibrating atoms in the crystal cause some electrons to break free. More thermal electrons are freed as the temperature rises. Every electron that becomes free leaves a gap or hole in an atom. Since the atoms are normally neutral the hole behaves as if it has a positive charge. Like electrons these positive holes seem to move through the semiconductor material and form part of the electric current in it. Conduction in a semiconductor by means of thermal electrons and positive holes is called intrinsic conduction.

A thermistor is a piece of semiconductor material behaving in this way. When cold the thermistor has a high resistance. As its temperature increases more electrons and holes are released, improving the intrinsic conduction of the material and hence lowering its resistance. Thermistors are used to prevent large currents flowing through lamp filaments and electric motors at the moment of switching on. They can also be used to operate thermostats and fire alarms.

Cadmium sulphide is one of several semiconductor materials whose resistance varies with the amount of light falling on it. As light energy is absorbed, electrons and holes are released and become available for conduction. The brighter the light the more of these charge carriers are released and the further the resistance of the material falls. Such semiconductor materials are used in light-dependent resistors (LDR).

Small quantities of different elements can be added to a semiconductor material which greatly change its conduction properties. Conduction caused by these impurities is called extrinsic conduction or impurity conduction. Adding such impurities to semiconductor materials is called doping. Silicon atoms have four electrons in their outer shells and all are required for bonding with its four nearest neighbours. If some atoms with five electrons in their outer shells are added to the silicon crystal, they provide one spare electron each. Phosphorus, antimony and arsenic are used to dope silicon in this way and provide spare electrons for conduction. Such doped material is called n-type silicon as it has an excess of negative charge carriers.

If materials such as aluminium, gallium and indium, with three electrons in the outer shells of their atoms, are used to dope the crystal, it will have one electron missing in its structure for each impurity atom present. Semiconductor material with extra positive hole charge carriers (missing electrons) is called p-type material.

## 24.7 The p–n junction diode

Such a diode consists of a single crystal of silicon or germanium, part of which has been doped so that it contains an excess of positive charges (p-type). The rest of the crystal has had a different impurity added so that it contains an excess of electrons (n-type). It is the existence of the junction between the two types of material which enables the device to act as a rectifier. It is therefore called a p–n junction diode.

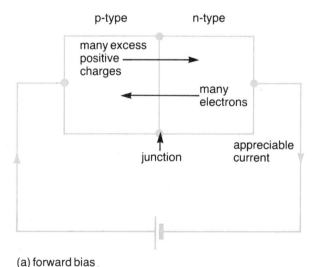

Fig 24.7

Suppose that a battery is connected across a junction diode as in Fig. 24.7(a), the so-called forward bias connection. The battery urges the excess positive charges and the electrons to cross the junction and so constitute a current. When the battery is reversed the excess charges are discouraged from crossing the junction and no current flows. This one way conduction property of a junction diode may be used to convert alternating current to direct current in the way described for the thermionic diode in Unit 24.2.

## 24.8 The light emitting diode (LED)

Junction diodes made of gallium arsenide and gallium phosphide emit light when a forward biased current flows through them. The colour of the light depends on the semiconductor material. The main advantage of the light emitting diode, compared to the filament lamp, is the low current and hence the low heat production. To limit the current to a low and safe value for the LED a protective resistor is connected in series with the diode (Fig. 24.8).

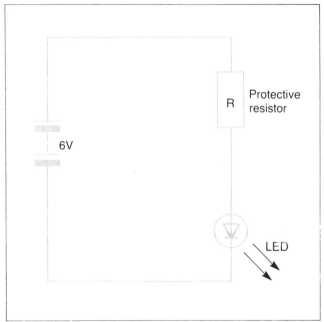

**Fig 24.8**

## 24.9 The transistor

The transistor is a semiconductor sandwich usually made of silicon. A thin layer of p-type silicon sandwiched between two layers of n-type silicon form an npn transistor. A pnp transistor has the types of silicon reversed.

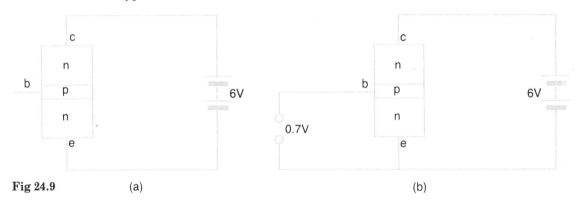

**Fig 24.9** (a) (b)

Consider an npn transistor, shown diagramatically in Fig. 24.9. The n-type regions are the emitter and collector; the thin p-type region is the base. If a potential difference is applied between the emitter and collector as shown in Fig. 24.9(a), the upper junction is reverse biased and no current flows through the transistor. If the lower junction is now forward biased by applying a small potential difference across it, as in Fig. 24.9(b), electrons flow from the emitter into the base. The base is so thin that most of these electrons drift across into the collector; the top junction is conducting and a current flows through the transistor as a result. A small current in the base or input circuit enables a much larger current to flow round the collector or output circuit. These results mean that:

1 The transistor can be used as a switch, because no current flows in the collector circuit unless a current flows in the base circuit.

2 The transistor can be used to amplify current changes as a small change in the base current produces a large change in the collector current.

In Fig. 24.9(b) note that the base and collector circuits share a common connection at the emitter. Used in this way the transistor is said to be in the common-emitter mode.

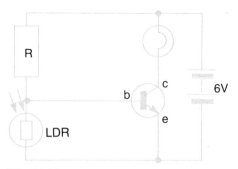

**Fig 24.10**

Figure 24.10 shows an npn transistor being used as a light operated transistor switch. When light falls on the LDR its resistance is low, the input voltage to the base is below about 0.7 V and the transistor is switched off. No current flows in the collector circuit and the lamp is off. When the LDR is shielded from the light its resistance rises. The input voltage to the base also rises above about 0.7 V causing current to flow in the base circuit. A larger current flows in the collector circuit and the lamp comes on. The value of the resistor $R$ is determined by the type of transistor and LDR used.

Figure 24.11 shows the transistor being used as a current amplifier. Small variations in the base current, caused by the microphone, lead to much greater variations in the collector current. The current amplification or gain of a transistor is given by the formula:

$$\text{current gain} = \frac{\text{collector current}}{\text{base current}}$$

The value of the current gain of a transistor can be anywhere between 10 and 1000 with a typical value of about 100.

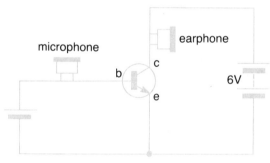

**Fig 24.11**

## 24.10 Electronic systems

Electronics systems are becoming increasingly more common and more important as time goes on. These systems can be described and understood in terms of building blocks and block diagrams. They handle and process two different kinds of information; analogue and digital. Here we confine our attention to digital systems.

In a digital system all information is processed in the form of digits; for example 1988 has four digits and each digit tells us how many thousands, hundreds, tens and ones of years there are. Calculators and computers represent numbers by binary digits. There are only two digits, 0 and 1, in binary arithmetic, rather than the ten digits 0 to 9 that we are used to in our usual arithmetic. For example the number 6 becomes 110 when expressed in binary digits. 110 stands for one four, one two and zero ones. Each of the binary digits 110 is called a bit and an electronic system which handles information in this way is a binary digital system. Note that calculators and computers convert the binary numbers into normal arithmetic before displaying the answer.

## 24.11 Logic gates

The use of a transistor as a switch has already been described. A variety of switches or building blocks called logic gates are used in digital systems. These blocks are like 'gates' because they have to be opened to let information pass through to reach their outputs. Each type of gate is opened by a particular combination of information fed to its inputs in binary code.

There are five basic types of logic gate from which all the more complicated ones are constructed. They are shown in Fig. 24.12. A truth table is a very simple way of describing all

| Type | Symbol | Same as | Truth table | Output is high (1) when: |
|------|--------|---------|-------------|--------------------------|
| NOT | A  —▷o— Y | INVERTER | input A: 0, 1 / output Y: 1, 0 | input A is NOT high |
| OR | A, B  —▷— Y | | A B Y: 0 0 0, 0 1 1, 1 0 1, 1 1 1 | input A OR B is high |
| NOR | A, B  —▷o— Y | OR–NOT | A B Y: 0 0 1, 0 1 0, 1 0 0, 1 1 0 | neither input A NOR input B is high |
| AND | A, B —D— Y | | A B Y: 0 0 0, 0 1 0, 1 0 0, 1 1 1 | input A AND input B are high |
| NAND | A, B —Do— Y | AND–NOT | A B Y: 0 0 1, 0 1 1, 1 0 1, 1 1 0 | Input A AND input B are NOT both high |

**Fig 24.12**

the possible combinations of inputs and outputs produced by a particular gate, or collection of gates. The name of a particular gate describes how it makes its decisions. It tells us which combination of high or NOT high (low) inputs produces a high output signal. Logic 0 represents a low input or output and logic 1 a high one. For example an AND gate only gives a high output (1) when both inputs are high (1).

Now, rather than manufacturing separate circuits containing individual transistors, integrated circuits (ICs) are used. Each integrated circuit consists of a very small single chip of silicon on which groups of components, including several transistors, are manufactured.

## 24.12 Summary

1. When a wire is heated electrons are released from the surface.
2. A thermionic diode contains a wire which is heated electrically (the cathode) and a metal plate (or cylinder) called the anode which is positive with respect to the cathode. Electrons released by the cathode are accelerated to the anode. Electrons cannot be made to flow in the opposite direction.
3. The diode (or electron gun) is the source of the electron beam in a Maltese cross tube, a deflection tube and a cathode ray oscilloscope.
4. Materials, such as metals, which readily allow electrons to flow through them are called conductors. Materials which do not are called insulators. A few materials, such as silicon and germanium have conducting properties between the two groups and are called semiconductors.
5. Some materials, cadmium sulphide, for example, have less resistance when light falls on them. Such materials are used in light dependent resistors (LDR).
6. A diode can be made from, for example, a single crystal of silicon or germanium which has been doped. Such a diode (junction diode) behaves in a circuit in a similar way to a thermionic diode.
7. Junction diodes made of gallium arsenide or phosphide emit light when a forward biased current flows through them. They are called light emitting diodes (LED).
8. A transistor is a semiconductor sandwich usually made from silicon. A thin layer of p-type silicon sandwiched between two layers of n-type silicon form an npn transistor. A pnp transistor has the types of silicon reversed.
9. A transistor is often used as a switch or to amplify a current.
10. Digital electronic systems use binary arithmetic and are the basis of calculators and computers.
11. There are many types of logic gate. Each type of gate is 'opened' by a particular combination of information fed to its inputs in binary code.

# 25 RADIOACTIVITY AND ATOMIC STRUCTURE

It has already been seen that electrons passing through a gas ionize the molecules of the gas (for example in the fine beam tube — Unit 24.4). Due to the electrical forces of attraction and repulsion, any charged particle passing near to a molecule will tend to ionize it — that is, split it into positively and negatively charged parts. This property of charged particles may be used to detect their presence.

## 25.1 Radiation detectors

One of the most widely used detectors is the Geiger-Müller (GM) tube shown in Fig. 25.1.

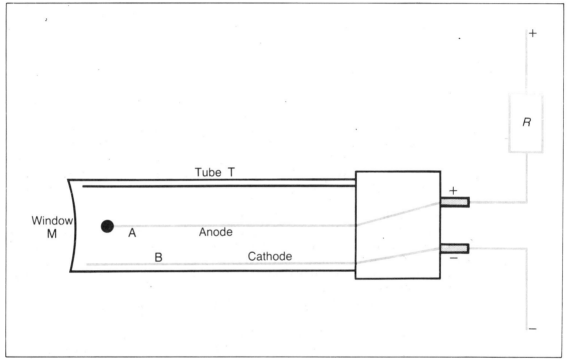

**Fig 25.1** Geiger tube

It consists of a small closed glass tube $T$, with a thin mica end window $M$, and contains a gas such as argon at about half atmospheric pressure. A thin wire $A$, which acts as the anode, passes down the centre of the tube and is insulated from it. The inside of the tube is coated with a conductor and forms the second electrode $B$. A potential difference of about 450 V is applied between the electrodes.

When a charged particle enters the tube through the thin mica window some argon atoms are ionized. Many more atoms throughout the tube then become ionized. The negative ions produced are attracted towards the central wire $A$, the positive ions going towards $B$. A current is thus obtained in the circuit for a short time. It is called a pulse of current and causes a voltage pulse in the high resistance $R$. The 450 V power supply and the resistance $R$ are contained in an electronic unit such as a scaler, which counts individual pulses, or a ratemeter, which gives the average count rate.

The presence of X-rays and $\gamma$-rays may also be detected using a GM tube.

A cloud chamber may also be used to detect the presence of charged particles. One type — the diffusion cloud chamber — is shown in Fig. 25.2.

It consists of a cylindrical perspex chamber with a detachable lid. The base of the chamber may be cooled by placing dry ice in a container below it. The felt just below the lid of the chamber is first moistened with methyl alcohol. The base of the container is then removed and dry ice packed under the base of the chamber. The sponge is replaced to keep the dry ice in contact with the base of the chamber. It is important that the chamber is levelled, thus reducing convection currents.

The dry ice cools the base of the chamber to about $-60°C$. Alcohol evaporates continuously from the felt below the lid, which is at room temperature, and the vapour sinks continuously to

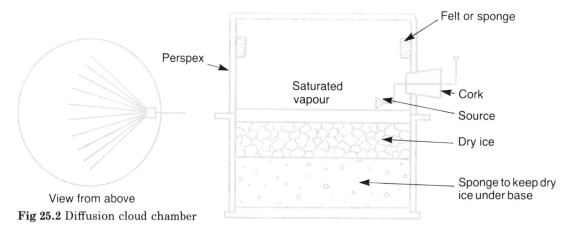

**Fig 25.2** Diffusion cloud chamber

the bottom of the chamber. Just above the cold base the vapour is supersaturated, that is, its vapour pressure is above the saturated vapour pressure at that temperature. When a charged particle passes through the chamber it ionizes some air molecules leaving a trail of positive and negative ions. These ions act as centres for condensation of the supersaturated vapour near the base of the chamber, and the path of the particle is revealed by the droplets which form.

## 25.2 Atomic structure

Atoms may be regarded as being like miniature solar systems. An atom is thought to have a central nucleus, consisting of tightly packed particles called protons and neutrons, with electrons revolving round it. Evidence for this model comes from a series of experiments carried out by Geiger and Marsden at Manchester University in 1911. Alpha particles emitted from a radioactive source were scattered by a thin foil of gold. A very small fraction of the particles were scattered through very large angles (some through more than 90°) and Rutherford suggested that these had come very close to a high concentration of positive charge (an atomic nucleus) and had been deflected by the large electrical force of repulsion.

In the nucleus of an atom each proton has a positive charge equal in magnitude to the charge on an electron, and the number of protons in the nucleus of an atom is equal to the number of electrons in the atom. The atom is therefore electrically neutral. The chemical properties of an atom and hence the element to which it belongs are dictated by the number and arrangement of electrons in it, and this in turn depends on the number of protons in the nucleus, known as the **atomic number**, denoted by $Z$.

Neutrons are similar in mass to protons, both being nearly 2000 times as massive as electrons. The mass of an atom is thus almost entirely due to the masses of the protons and neutrons contained in its nucleus. The total number of protons and neutrons in the nucleus of an atom is known as the atom's **mass number**, denoted by $A$. Thus if the number of neutrons is denoted by $N$ we have

$$A = Z + N$$

It is possible for one or more of the peripheral electrons of an atom to become detached, thus leaving the atom with a net positive charge. It is then known as a positive ion. It is also possible for an atom to gain one or more electrons, becoming a negative ion. It is only in radioactive disintegrations that the number of protons or neutrons in an atom changes. All the electrical and chemical properties of an atom are explained in terms of the transfer of electrons.

Helium has a nucleus containing two protons and two neutrons, and two peripheral electrons. When an atom disintegrates by means of **alpha particle decay** (see Unit 25.4) it loses two protons and becomes an atom of the element two below it in the periodic table of elements. As it has also lost two neutrons, its mass number has fallen by four.

**Beta decay** seems to take place by a neutron changing to a proton, which remains in the nucleus, and an electron which is emitted. The atom concerned has gained a proton and thus becomes an atom of the element one up in the periodic table. Its mass number has not changed.

## 25.3 Isotopes

From the previous discussion it can be seen that two atoms of an element may exist (that is the atoms have the same number of protons) which have different numbers of neutrons and hence different mass numbers. These atoms are said to be **isotopes of the same element**.

At the lower end of the periodic table the number of neutrons present in an atom is equal to, or nearly equal to, the number of protons. Elements of high mass number consist of atoms with very many more neutrons than protons. It is these atoms which are more likely to disintegrate.

It seems that a large number of excess neutrons in an atom leads to instability and the likelihood of spontaneous disintegration. Generally speaking it is the elements of high mass number which exhibit this property of radioactivity.

## 25.4 Radioactivity

A number of naturally occurring substances emit particles or radiations which ionize gases. Marie Curie and her husband did much of the early work on radioactive substances and showed that amongst the most active were those containing uranium, polonium and radium.

By 1899 Rutherford had shown that the particles or radiations emitted by radioactive substances fell into three categories which he called **alpha ($\alpha$), beta ($\beta$)** and **gamma ($\gamma$) rays**.

**Alpha rays** are helium ions — that is, helium atoms which have lost two electrons, and hence have a positive charge. From a particular radioactive substance they are all ejected with approximately the same velocity and hence kinetic energy. They have a range of a few centimetres in air at atmospheric pressure, but most are stopped by a thick sheet of paper. Like all charged particles they lose their energy by continuous ionization and the fact that they produce many ions per centimetre of path means that they travel relatively short distances. Being charged particles they are deflected by both electric and magnetic fields.

**Beta rays** are streams of high energy electrons similar to cathode rays but travelling much faster. They are emitted with velocities approaching that of light ($3 \times 10^8$ m/s) and, as they do not form such a high density of ions along their track, their range is greater than that of alpha rays. Beta rays are negatively charged and are thus deflected in the opposite direction to alpha particles in both electric and magnetic fields.

**Gamma rays** are very short wavelength radiations similar to X-rays, which can penetrate several centimetres of dense metal such as lead. They do not produce continuous ionization but lose their energy in one interaction with a molecule. The fact that they cannot be deflected by electric or magnetic fields indicates that they are not charged particles.

An idea of the range of alpha, beta and gamma rays may be obtained by placing sources emitting these separate rays in front of a GM tube connected to a scaler. The number of counts recorded by the scaler in a given time is noted with different absorbers (sheets of paper, aluminium or lead of different thicknesses) between the source and the tube. Alphas are stopped by a thin sheet of paper. It is found that gamma rays in particular and beta rays to a lesser extent are very difficult to stop or screen, and thus may present a safety hazard. Gamma ray sources are kept in lead containers, the walls of which are often several centimetres thick.

If the GM tube and scaler are switched on in the absence of a source and the absorbers, some counts will be recorded. This count rate, which is slow, is due to radiation which is always present in the atmosphere. It is called background radiation, and should be subtracted from the counts recorded, when a source is present, to give the true count rate due to the source.

## 25.5 Radioactive decay

By 1903 Rutherford had come to the conclusion that radioactivity is the result of the spontaneous disintegration of an atom during which it emits an alpha or beta ray. Simultaneously the atom changes into one of another element which may itself be radioactive. The following are examples of such disintegrations.

$^{238}_{92}U \rightarrow ^{4}_{2}He + ^{234}_{90}Th$    $\alpha$-decay

$^{234}_{90}Th \rightarrow ^{0}_{-1}e + ^{234}_{91}Pa$    $\beta$-decay

$^{234}_{91}Pa \rightarrow ^{0}_{-1}e + ^{234}_{92}U$    $\beta$-decay

The numbers above the symbols are the mass numbers of the atoms; those below are the atomic numbers.

It is impossible to predict which particular atom in a sample will change in this way; that is, the disintegration of an atom is a random event. However, in a sample containing a large number of atoms, the number which disintegrate each second will depend on the number of undecayed atoms left. Thus if the number of disintegrations per second is recorded (by counting the number of particles emitted) and plotted against time an exponential graph results (Fig. 25.3).

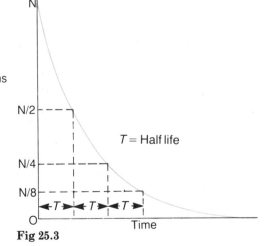

Fig 25.3

As can be seen from this graph, an infinite time is taken for all the atoms of a sample to disintegrate. It is therefore meaningless to talk about the 'life' of a sample. However, **the time taken for half the undecayed atoms in any given sample of the substance to disintegrate is finite, and the same whatever the number of undecayed atoms present initially**. This time $T$ is known as the **half-life** of the substance. It is different for different radioactive materials.

## 25.6 Safety

Exposure of the body to radiation from atomic disintegrations can have undesirable effects of a long- or short-term nature. The precise result of exposure depends on the nature of the radiation, the part of the body irradiated and the dose received. The hazard from alpha particles is slight, unless the source enters the body, since they cannot penetrate the outer layers of skin. Beta particles are more penetrating, although most of their energy is absorbed by surface tissues. Gamma rays present the main external radiation hazard since they penetrate deeply into the body.

Radiation can cause immediate damage to cells, and is accompanied by radiation burns (redness of skin followed by blistering and sores, the severity of these depending on the dose received), radiation sickness and possibly death. Effects such as cancer and leukemia may appear many years later, due to the uncontrolled multiplication of some cells set off by exposure to radiation. Hereditary effects may also occur in succeeding generations due to genetic changes. The most susceptible parts of the body are the reproductive organs and blood-forming organs such as the liver.

Because of the hazards it is essential that the correct procedure be adopted when using radioactive substances.

Briefly it is as follows:

1. Sources should only be held with forceps, never with the hand. This avoids any possibility of some of the substance transferring to the surface of the skin or lodging under a nail.
2. Any cuts on the hand should be covered before using sources.
3. Sources should not be pointed towards the human body.
4. Sources should be returned to their container as soon as they are no longer needed for an experiment. If possible, permanent storage should be within two containers.
5. A check should be made at the end of an experiment to see that all sources are present in their allotted containers.
6. Mouth pipettes should never be used with liquid sources.
7. Do not remain unnecessarily in the region of a radioactive source.

## 25.7 Nuclear energy

The nucleus of a $^{235}_{92}U$ atom splits into two parts with approximately equal masses when it captures a neutron. The splitting of the nucleus is known as nuclear fission. The products of the fission of one uranium-235 nucleus are:

1. Two new elements, known as fission products. A pair of examples is barium and krypton.
2. Two or three fast-moving neutrons.
3. About $3 \times 10^{11}$ J of energy. This energy is released as radiation and as kinetic energy of the neutrons. This kinetic energy is the source of the heat energy obtained from a nuclear reactor.

$$^{235}_{92}U + ^{1}_{0}n \rightarrow ^{144}_{56}Ba + ^{90}_{36}Kr + 2^{1}_{0}n$$

The nuclear fission equation above shows that the number of nuclear particles (nucleons) is conserved. However, the total mass of the particles on the right-hand side of the equation is slightly less than the total mass on the left. This small mass difference is converted into energy according to the equation:

$$E = mc^2$$

where $E$ is the energy released, $m$ is the mass difference and $c$ is the speed of light ($c = 3 \times 10^8$ m/s).

The fission of one uranium-235 nucleus gives off about $3 \times 10^{11}$ J of energy, which is far more than that released by natural radioactive decay.

If the neutrons released by the decay of one uranium-235 nucleus are captured by other nearby uranium-235 nuclei then more fissions occur, resulting in a chain reaction. About 85 per cent of the uranium-235 nuclei which capture a neutron undergo fission, emitting two or three more neutrons. If all these neutrons were captured by uranium-235 nuclei a nuclear explosion would result. This does not happen because in a nuclear reactor it is possible to control the chain reaction. Natural uranium contains only about 0.7 per cent of the isotope $^{235}_{92}U$. 99.3 per cent of the atoms present are of the isotope $^{238}_{92}U$ which captures neutrons without fission resulting. Thus in natural uranium the chain reaction would quickly stop.

However, the probability of uranium-235 nuclei capturing neutrons can be increased by slowing the neutrons down. This also reduces the probability of capture by uranium-238 nuclei. The neutrons are slowed down by a material called a moderator, usually water or graphite. Neutrons which have been slowed down are known as thermal neutrons, because they have kinetic energies similar in value to the thermal (heat) energy of the surrounding material.

If the fission process is to be used in a power station, it is necessary to control the chain reaction so that exactly one neutron from the fission of each uranium-235 nucleus causes another similar nucleus to split. If, on average, less than one neutron does so, then the chain reaction will not be self-sustaining. If the number is above one the chain reaction will quickly get out of control and an explosion will result. This control is achieved by the use of control rods of boron steel or cadmium, which absorb neutrons. The length of the control rods in the uranium core is continually adjusted automatically so that an average of one neutron per nuclear fission causes further splitting.

The heat produced by fission is removed from the reactor core by a coolant which is piped into the core. The pipe containing the coolant passes close to the hot uranium fuel rods and the moderator and much of the heat is absorbed. The heated coolant is then used to run a turbine. From this point onwards a nuclear power station works in the same way as power stations burning oil or coal.

## 25.8 Summary

1. Detectors of radiation rely on the fact that emissions from radioactive sources (e.g. alpha, beta and gamma emissions) all ionize some molecules of any gas through which they pass.
2. An atom consists of a central very small nucleus which contains all the positive charge and most of the mass of the atom. The nucleus is surrounded by electrons (negative charge) in motion. An atom contains an equal number of positive and negative charges.
3. When a substance decays it gives out alpha, beta or gamma radiation or a combination of these.
4. Alpha particles are helium nuclei, beta particles are electrons, and gamma rays are high frequency electromagnetic waves.
5. The half life of a radioactive substance is the time taken for half the undecayed atoms to decay.
6. It is vital that correct safety procedures are followed when using radioactive materials.
7. The nucleus of a $^{235}_{92}U$ atom splits into two parts of approximately equal masses when it captures a neutron. In addition, 2 or 3 fast moving neutrons and about $3 \times 10^{11}$ J of energy are released.
8. If one or more of the neutrons released by the splitting of each $^{235}_{92}U$ nucleus are captured by other $^{235}_{92}U$ nuclei, more splitting (fission) occurs.
9. If on average one of the neutrons released in the splitting of each nucleus causes further splitting then a controlled chain reaction results.
10. There are few $^{235}_{92}U$ atoms in naturally occurring uranium. More than 99 per cent of natural uranium is $^{238}_{92}U$ atoms which tend to capture neutrons without fission resulting.

# SELF-TEST QUESTIONS

This section contains a selection of multiple choice questions and is designed to test your knowledge of the material used in the book. Before attempting these questions you must read the unit entitled 'Hints for candidates taking Physics examinations' (pages 119–120).

## Questions (answers p 118)

### Type 1

Each question below has five possible answers labelled A, B, C, D, E. For each numbered question select the one which is the best answer. Within the group of questions each answer may be used once, more than once or not at all.

*Questions 1–3*

    A newton per second         D newton metre
    B kilogram metre per second   E kilogram metre per second$^2$
    C newton metre per second

Which one of the above is a unit in which each of the following physical quantities might be measured?

1 momentum

2 work

3 power

*Questions 4–7*

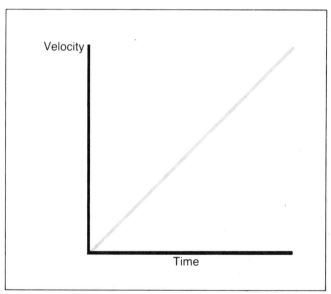

The graph shows how the velocity of a moving object starting from rest changes with time.

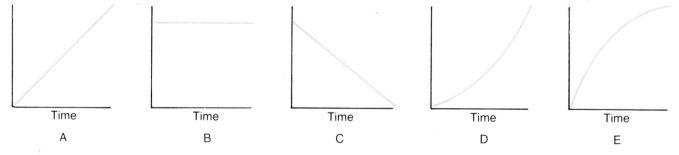

Select from the graphs A–E above the one which most nearly represents the shape of each of the following graphs for the moving object.

4 acceleration against time
5 the force acting on the body against time
6 distance travelled against time
7 kinetic energy against time

*Questions 8–12*
The following are five properties of a body made of the same metal throughout.
   A mass   B volume   C weight   D density   E surface area
Which of the properties

8 is a vector quantity?
9 would change if the body were taken to the moon, its temperature remaining constant?
10 is calculated from two of the other properties?
11 would not change if the body were cut in half and only one half were considered?
12 has units of kilogram metre per second$^2$?

*Questions 13–15*
The following are five units
   A kilogram                     D newton
   B kilogram metre per second   E metre per second$^2$
   C joule
Which of the above units is suitable for measuring each of the following quantities?

13 the kinetic energy of a cricket ball
14 the momentum of a cricket ball
15 the weight of a cricket ball

*Questions 16–18*
Which of the following words best describes the particles in question?
   A protons   B molecules   C ions   D electrons   E neutrons

16 carbon dioxide gas at room temperature
17 particles emitted from a hot filament in the tube of a cathode ray oscilloscope
18 alpha particles emitted by a radioactive substance

*Questions 19–21*
Five types of radiation are listed below.
   A ultraviolet radiation   D $\beta$-radiation
   B infrared radiation      E ultrasonic radiation
   C $\gamma$-radiation
Select the type of radiation to which each of the following applies.

19 The wavelength is slightly shorter than that of visible light and it can cause certain materials to fluoresce.
20 The wavelength is extremely short, much shorter than that of visible light, and it can be emitted from the nucleus of an atom.
21 The wavelength is slightly longer than that of visible light and it may be used for heating.

*Questions 22–24*
Resistors of 1 ohm, 1 ohm and 2 ohms are connected in arrangements lettered A to E below.

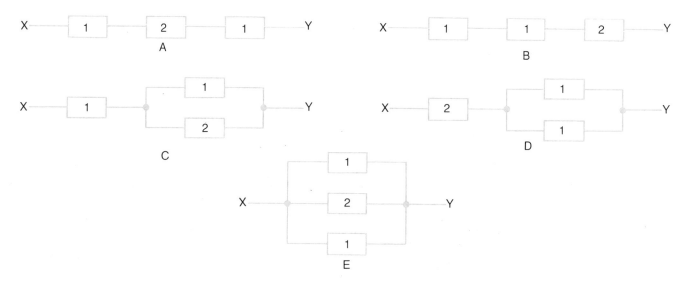

22 Which arrangement has the smallest total resistance between X and Y?

23 Which arrangement has a total resistance of 2.5 ohms between X and Y?

24 Which arrangement would have the largest current through the 2 ohm resistor when a potential difference is applied between X and Y?

**Type 2**

Each of the following questions or incomplete statements is followed by five suggested answers labelled A, B, C, D, E. Select the best answer in each case.

25 Which of the following is measured in newtons?
   A energy   B force   C power   D pressure   E work

26 Which of the graphs best represents the velocity–time graph of a ball thrown vertically upwards to a considerable height and which then returns to the thrower?

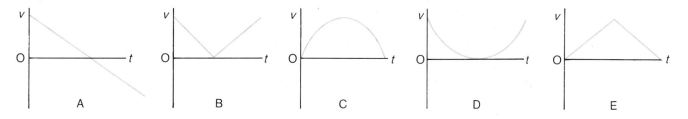

*Questions 27–28*
A car takes 5 seconds to accelerate uniformly from rest until it reaches a speed of 10 metres per second.

27 What is its acceleration?
   A $0.5 \text{ m/s}^2$   B $2 \text{ m/s}^2$   C $5 \text{ m/s}^2$   D $10 \text{ m/s}^2$   D $50 \text{ m/s}^2$

28 How far does it travel during the 5 seconds?
   A 2 m   B 5 m   C 10 m   D 25 m   E 50 m

29 The force acting on a moving object is constant in magnitude and is always perpendicular to the direction of motion of the object. How is the object moving?
   A accelerating with uniform acceleration in a straight line
   B decelerating with uniform deceleration in a straight line
   C moving in an elliptical path
   D continuing in a straight line
   E moving in a circular path

30 A 5 kg mass is travelling with a speed of 5 m/s. It is brought to rest in 0.5 s. What is the average force acting on it to bring it to rest?
   A 50 N   B 25 N   C 10 N   D 2.5 N   E 0.5 N

31 An astronaut goes to the moon where the gravitational attraction is less than on the earth. Which one of the following correctly describes the change(s) in his mass and/or weight?

    *Mass*     *Weight*
A increases   unchanged
B unchanged   unchanged
C unchanged   decreases
D decreases   unchanged
E decreases   decreases

32 A car of mass 800 kg which is moving east at 50 m/s collides head on with a van of mass 1600 kg which is moving west at 10 m/s. What will the wreck do if the vehicles stick together?

A be stationary     D move east at 10 m/s
B move east at 20 m/s   E move west at 10 m/s
C move west at 20 m/s

33 8 kg of water is tipped from a bucket at the top of a building of height 200 m and forms a static pool on the ground. What will be its change in energy?

A 1 600 000 J gain   D 16 000 J loss
B 16 000 J gain     E 1 600 000 J loss
C no change

34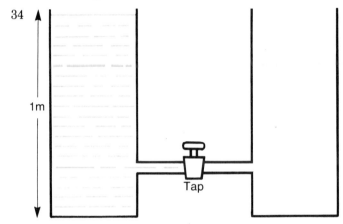

The tank in the diagram is filled to a depth of 1 m with 10 kg of water. The tank is joined to an identical tank by a pipe with a tap in it. When the tap is opened each of the tanks becomes half-full of water. How much potential energy is transformed when this happens?

A 0   B 2.5 J   C 5 J   D 10 J   E 25 J

35 Which of the following is the unit of work?

A watt     D newton/metre
B joule     E metre
C newton

36 A ship is sailing slowly due east, and a passenger is on the deck at *P*. In the diagram the points *A*, *B*, *C*, *D* and *E* are marked on the deck. Towards which of these points should the passenger walk at a suitable speed in order to move due north?

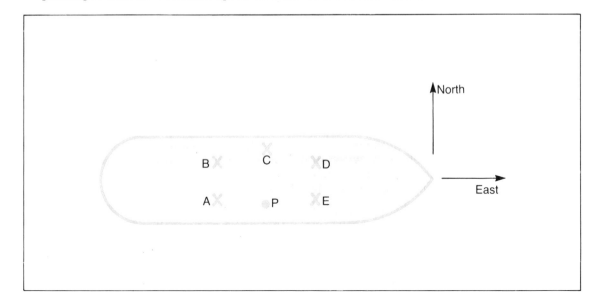

*Questions 37 and 38*
The diagram below shows a uniform beam *XY* of length 1 m supported at *P* where *PX* = 20 cm. The beam weighs 5 N.

37 What downward force exerted at *X* will make the beam balance?

A 5 N    B 7.5 N    C 12.5 N    D 20 N    E 25 N

38 What is the value of the force *F* if the beam shown in the diagram is balanced?

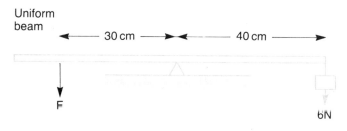

A 3 N    B 4.5 N    C 6 N    D 8 N    E 10 N

39 A uniform rod *AB* is 2 m long and weighs 3 N. It is rested on a pivot *P*, 0.5 m from *A*. How should a force of 3 N be applied to balance the rod?

A upwards at *A*
B downwards at *A*
C upwards at *B*
D downwards at *B*
E downwards at the centre of mass of the rod

40 Where does the centre of mass of a triangular shaped piece of thin cardboard lie?
   A on a line from one vertex perpendicular to the opposite side
   B on a line through the mid-point of one side and perpendicular to that side
   C on a line through the mid-point of one side and through the opposite vertex
   D on a line through the mid-point of one side and parallel to an adjacent side
   E on a line bisecting the angle at a vertex

41 A single string pulley system is illustrated below. A load of 10 newtons can be raised by an effort of 4 newtons.
What is the weight of the lower pulley block if the system is frictionless?

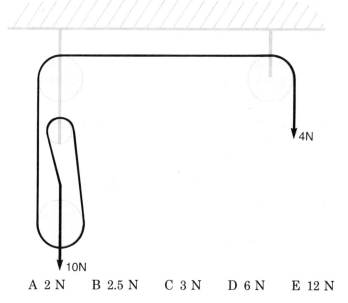

A 2 N    B 2.5 N    C 3 N    D 6 N    E 12 N

42 What is the volume of a piece of iron of mass 40 kg and density $8 \times 10^3$ kg/m$^3$?
A $32 \times 10^4$ m$^3$   B 200 m$^3$   C 20 m$^3$   D $5 \times 10^{-2}$ m$^3$   E $5 \times 10^{-3}$ m$^3$

43 A metal cube, with each side 2 cm in length, has a density of 8 g/cm$^3$. What is the mass of the cube?
A 2 g   B 4 g   C 8 g   D 16 g   E 64 g

44 A boy of mass 40 kg balances evenly on two stilts, each having an area of 8 cm$^2$ in contact with the ground. What is the pressure exerted by one stilt?
A 50 N/cm   B 40 N/cm   C 25 N/cm   D 5 N/cm   E 2.5 N/cm

45 When a force $F$ acts on an area $A$, the pressure is $P$. What is the pressure when a force $2F$ acts on an area $\frac{1}{2}A$?
A $4P$   B $2P$   C $P$   D $\frac{1}{2}P$   E $\frac{1}{4}P$

46 A car's tyres are inflated so that an area of 50 cm$^2$ of each tyre is in contact with the ground. If the mass of the car is 1600 kg and is evenly distributed between the four tyres, what is the pressure in each tyre?
A 8 N/cm$^2$   B 32 N/cm$^2$   C 80 N/cm$^2$   D 160 N/cm$^2$   E 320 N/cm$^2$

47 The block shown has a weight of 120 N. What pressure would it exert if stood on its smallest face?

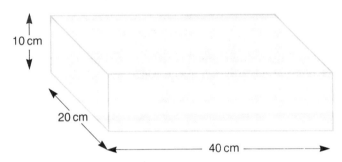

A 0.15 N/cm$^2$   B 0.30 N/cm$^2$   C 0.60 N/cm$^2$   D 1.20 N/cm$^2$   E 2.40 N/cm$^2$

48 A spring has an unstretched length of 100 mm. When a force of 5 N is applied the new length of the spring is 150 mm. Assuming that the elastic limit is not exceeded, what is the length of the spring when a force of 7.5 N is applied?
A 75 mm   B 125 mm   C 150 mm   D 175 mm   E 225 mm

49 When a mass of 300 g is hung on a spring, the length of the spring increases by 3.0 cm. By how much would it increase in length if a mass of 200 g were hung on it?
A 2.0 cm   B 3.0 cm   C 4.0 cm   D 4.5 cm   E 6.0 cm

50 When viewed through a low power microscope, smoke particles in an air cell are seen to be in continuous random motion. What causes this motion?
A smoke particles colliding with smoke particles   D convection currents in the air
B air molecules colliding with smoke particles   E none of these
C air molecules colliding with air molecules

51 Which one of the following statements about the molecules of a substance is true?
A In a liquid at a constant temperature they all have the same speed.
B In a gas at a constant temperature a particular molecule has a fixed speed.
C In a liquid they are very much further apart than in a solid.
D In a solid the molecules will have more kinetic energy at a higher temperature.
E When a liquid boils only those near the surface escape into the surroundings.

52 A bar of copper is heated from 10°C to 20°C. Which of the following statements is NOT true?
A Its length will increase slightly.   D Its mass will remain unchanged.
B Its electrical resistance will increase slightly.   E Its weight will remain unchanged.
C Its density will increase slightly.

53 In an uncalibrated thermometer the length of mercury in the stem is 4 cm when the thermometer is in pure melting ice, and 44 cm when in pure steam, both under standard pressure. When placed in a liquid of unknown temperature, the length is 48 cm. What is the temperature of the liquid?

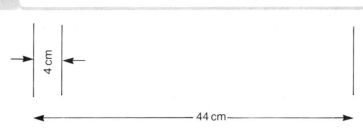

    A 48°C    B $109\frac{1}{11}$°C    C 110°C    D 140°C    E 160°C

54 A fixed mass of gas has its pressure halved and its temperature, on the Kelvin scale, doubled. What is the ratio of the new volume to the original volume?
    A 1:4    B 1:2    C 1:1    D 2:1    E 4:1

55 An electric heating wire is immersed in 0.05 kg of oil in a calorimeter of negligible heat capacity. The temperature of the oil rises from 20°C to 50°C in 100 seconds. If the specific heat capacity of oil is 2000 J/kg°C, what is the power supplied by the heating coil?
    A 20 W    B 30 W    C 50 W    D 3000 W    E 3 000 000 W

56 What is the SI unit of specific latent heat?
    A joule per kelvin    C joule per kilogram per kelvin    E joule per kilogram
    B joule per gram per kelvin    D joule per gram

57 How does heat travel through a vacuum?
    A by conduction only    D by conduction and convection
    B by convection only    E by conduction and radiation
    C by radiation only

58 In a ripple tank, waves travel a distance of 45 cm in 3 s. If the distance apart of the crests is 3 cm, what is the frequency of the vibrator causing the waves?
    A 5 Hz    B 7.5 Hz    C 11.25 Hz    D 20 Hz    E some other value

59 Which of the following describes the image formed in a pinhole camera by a distant upright object?
    A upright, real and magnified    D inverted, real and diminished
    B upright, real and diminished    E inverted, virtual and magnified
    C upright, virtual and diminished

60 Which of the following describes the image formed by a plane mirror?
    A real and the same size as the object
    B real and nearly the same size as the object
    C virtual and the same size as the object
    D virtual and nearly the same size as the object
    E virtual and half the size of the object

61 What happens when waves move from deep to shallow water?
    A Their wavelength decreases because their velocity decreases.
    B Their wavelength decreases because their velocity increases.
    C Their frequency increases because their velocity increases.
    D Their frequency decreases because their velocity increases.
    E Their frequency decreases because their velocity decreases.

62 Which one of the following must be moved in order to give the correct sequence of radiations in order of decreasing wavelength?
    A radio waves    B ultraviolet rays    C infrared rays    D yellow light    E gamma rays

63 In what way are sound waves different from light waves?
    A Sound waves need no medium, whereas light waves do.
    B Sound waves need a medium, whereas light waves do not.
    C Light waves travel through glass, whereas sound waves do not.
    D Light waves are reflected, whereas sound waves are not.
    E Sound waves are not refracted, whereas light waves are.

64 Which of the following statements about waves is true?
    A Radio waves, light waves and sound waves will all travel through a vacuum.
    B Radio waves and light waves will travel through a vacuum, sound waves will not.
    C Sound waves will travel through a vacuum, radio waves and light waves will not.
    D Light waves and sound waves will travel through a vacuum, radio waves will not.
    E None of the waves will travel through a vacuum.

65 A loudspeaker gives out a note of frequency 100 Hz. If the speed of sound is 330 m/s, what is the wavelength of the sound?
A 1.1 m    B 3.3 m    C 11 m    D 33 m    E 110 m

66 An unmagnetized bar of soft iron is tested at both ends by the south pole of a permanent magnet. Which of the following is observed?
A attraction to both ends
B attraction to one end only
C repulsion from both ends
D repulsion from one end only
E attraction to one end and repulsion from the other

67 A rod of insulating material is given a positive charge by rubbing it with a piece of fabric. The fabric is then tested for electric charge. What would you expect the fabric to have?
A a positive charge equal to that on the rod
B a negative charge equal to that on the rod
C a positive charge less than that on the rod
D a negative charge greater than that on the rod
E no charge

*Questions 68–69*
Two coils of wire, of resistance 2 ohms and 3 ohms, respectively, are connected in series with a 10 volt battery of negligible internal resistance as shown.

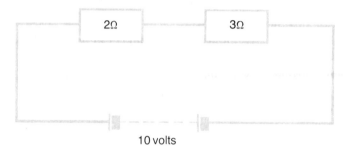

68 What is the current through the 2 ohm coil?
A 0.5 A    B 2 A    C 5 A    D 20 A    E 50 A

69 What is the potential difference across the 3 ohm coil?
A 2 V    B 4 V    C 5 V    D 6 V    E 10 V

*Questions 70–71*
Similar cells, ammeters and resistors are used in the circuit shown. The cells and ammeters have negligible resistance. When one cell is connected in series with one ammeter and one resistor the ammeter reading is 0.1 A.

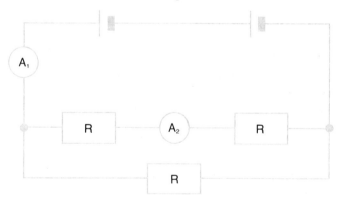

A 0.1 A    B 0.2 A    C 0.3 A    D 0.4 A    E 0.5 A

70 Select from the list above the value of the current through ammeter $A_1$.

71 Select from the list above the value of the current through ammeter $A_2$.

72 A 24 watt 12 volt headlamp is lit by connecting it to a 12 volt battery of negligible resistance. What is the current through the headlamp?
A 288 A    B 24 A    C 12 A    D 2 A    E 0.5 A

73 An electric fire is rated 250 V 1000 W. If electricity costs 6p per unit what is the cost of running the fire for 5 hours?
A 3p    B 6p    C 15p    D 30p    E 60p

74 Why is the outer casing of an electric iron generally connected to earth?
   A to prevent a serious electric shock     D to protect the iron
   B to complete the circuit                 E to allow the current to get away
   C to prevent the fuse from burning out

75 From which material are the cores of electromagnets usually made?
   A cast iron     D cobalt steel
   B soft iron     E nickel steel
   C carbon

76 A 12 V 36 W lamp is connected across the output of a transformer. The output has 60 turns round the transformer. Which one of the following inputs to the transformer would enable the lamp to glow with normal brightness?

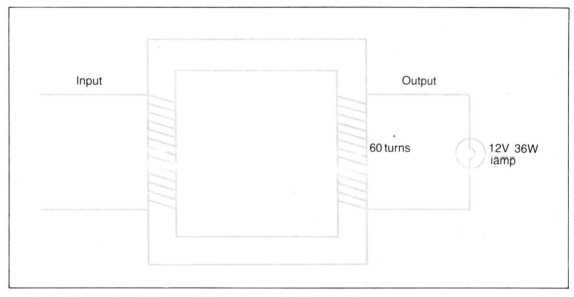

   A a voltage of 12 V d.c. and 36 turns on the input coil
   B a voltage of 120 V d.c. and 600 turns on the input coil
   C a voltage of 12 V a.c. and 36 turns on the input coil
   D a voltage of 120 V a.c. and 600 turns on the input coil
   E a voltage of 120 V a.c. and 6 turns on the input coil

*Questions 77–78*
In a school experiment a stream of electrons passes through a horizontal slit and strikes an inclined screen so that a trace is seen as indicated in the diagram.

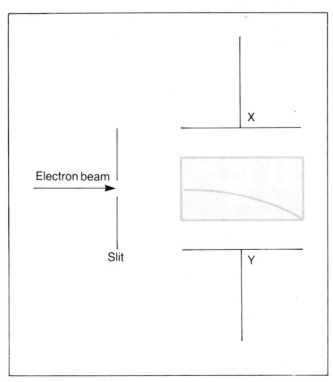

77 Which one of the following is the best explanation of the parabolic path?
   A The electrons are falling under the influence of gravity.
   B There is a magnetic field acting downwards between the plates.
   C The electrons are slowing down and losing energy.
   D Plate $X$ has a positive potential relative to plate $Y$.
   E Plate $X$ has a negative potential relative to plate $Y$.

78 Which one of the following procedures is the most likely to change the trace to a horizontal line?
   A increase the velocity of the electrons from the slit
   B apply a magnetic field in the direction of the beam
   C apply a magnetic field in the direction $XY$
   D apply a magnetic field in the direction $YX$
   E apply a magnetic field at right angles to both $XY$ and the electron stream

79 Atoms of atomic number (proton number) 92 and mass number (nucleon number) 234 decay to form new atoms of atomic number 90 and mass number 230. What will the emissions consist of?
   A electrons
   B neutrons
   C gamma rays
   D beta particles
   E alpha particles

80 Thorium 232 with atomic number 90 decays to element $X$ by ejecting an alpha-particle. Element $X$ emits a beta-particle to become element $Y$. Element $Y$ emits another beta-particle to become element $Z$. What are the mass number and the atomic number of element $Z$?

|   | Mass number | Atomic number |
|---|---|---|
| A | 230 | 90 |
| B | 228 | 88 |
| C | 228 | 90 |
| D | 230 | 88 |
| E | 226 | 90 |

## Answers to self-test units

| | | | | | | | |
|---|---|---|---|---|---|---|---|
| 1 B | 11 D | 21 B | 31 C | 41 A | 51 D | 61 A | 71 A |
| 2 D | 12 C | 22 E | 32 D | 42 E | 52 C | 62 B | 72 D |
| 3 C | 13 C | 23 D | 33 D | 43 E | 53 C | 63 C | 73 D |
| 4 B | 14 B | 24 E | 34 E | 44 C | 54 E | 64 B | 74 A |
| 5 B | 15 D | 25 B | 35 B | 45 A | 55 B | 65 B | 75 B |
| 6 D | 16 B | 26 A | 36 B | 46 C | 56 E | 66 A | 76 D |
| 7 D | 17 D | 27 B | 37 B | 47 C | 57 C | 67 B | 77 E |
| 8 C | 18 C | 28 D | 38 D | 48 D | 58 A | 68 B | 78 E |
| 9 C | 19 A | 29 E | 39 B | 49 A | 59 D | 69 D | 79 E |
| 10 D | 20 C | 30 A | 40 C | 50 B | 60 C | 70 C | 80 C |

# HINTS FOR CANDIDATES TAKING PHYSICS EXAMINATIONS

## Multiple choice or fixed response questions

All questions must be attempted in multiple choice, or as they are sometimes called 'fixed response', examination papers. It is therefore unnecessary and a waste of time to read the paper through before starting to write down your answers. However, as you come to each question do read it carefully before answering it.

Start at Question 1 and work steadily through the paper doing all the questions you can answer fairly easily. Do not delay, at this stage, over questions which you find difficult. Working in this way you should be able to reach the end of the paper in a little over half the time allowed for it, having answered about two-thirds of the questions. Now return to the beginning of the paper and tackle the questions you found difficult the first time through, but still leave out questions about which you have little or no idea. You should aim to reach the end of the paper five or ten minutes before the end of the examination.

Now is the time to return to the few questions still unanswered and make an *intelligent* guess in each case. Generally speaking you will be able to eliminate two or three of the alternative answers to each question, but you may not be sure which one of those left is correct. If you guess, at this stage, you will have a one in two, or one in three, chance of being right. However, *wild guessing is useless* and you should not guess at any answers until the last few minutes of the examination. *Do not leave any questions unanswered.* Examining Groups do not deduct marks for wrong answers.

If you have any time left, when you have answered all the questions, check through your answer sheet, bearing in mind the following points:

1 All questions should have an answer.
2 No question should have more than one answer — this can occur by mistake when filling in the sheet.
3 Any alteration should be such that the incorrect answer has been completely rubbed out.
4 A soft pencil should always be used for completing the answer sheet.

Questions on multiple choice examination papers fall into two basic types. Questions of the first type can be found under the heading **Type 1** in the self-test units (pages 109–111). The questions occur in groups with the same alternative answers for each question within the group, and the candidate has to select the most appropriate answer for each question. Within each group of questions each answer may be used once, more than once, or not at all. In **Type 2** each question has its own group of alternative answers and the candidate must select the best answer to each one (see pages 111–118).

## Free response questions

Free response questions can be defined as any question where the candidate has to make a written answer rather than just choose from the answers given. There is a wide range of free response questions ranging from short answers (where the answer may be a few words or sentences) through to longer structured questions.

### SHORT QUESTIONS AND STRUCTURED QUESTIONS

A selection of short and structured questions is given in the unit entitled 'Practice in answering examination questions' (pages 134–151). In these types of question space for the answers is often left on the question paper. If this is the case, the space left is a guide, but only a guide, to the length of the answer required. Do not feel that you *have* to fill the space, but if you find your answer is much too short or much too long think again. Whether space is left to answer on the question paper or not, your answers should contain all the relevant points stated in a logical order, and concisely expressed.

### LONGER QUESTIONS

A selection of longer questions is given. Examination papers containing this type of question vary considerably in their style and length of question, as well as how many questions they require the candidate to answer. It is essential to read the instructions printed at the beginning of the paper to find out which questions, if any, are compulsory, and how many of the others should be answered. Do not answer more questions than instructed.

If you are presented with a choice of question or parts of a question, it is essential that you read through *all* the alternatives before making your choice. It is not only important to make the correct choice for you, but then to do *first* that question or part question which you consider easiest. You are more likely to do well on the more difficult questions if you have already successfully completed some easier ones. However, be sure to number clearly the answers to questions and parts of questions.

It is also important that you divide up your time for the examination correctly. If, for example, four questions must be answered in two hours, then you cannot afford to spend more than 35 minutes on any one question. In fact the first question or two should each take you slightly less than half an hour rather than more.

Generally more time is allowed for answering each question than is required simply to write the answer down. It is therefore sensible to spend a few minutes planning your answer. First read the question carefully, then think which physical principles are involved. If all or part of the question requires a descriptive answer, it is worth jotting down in rough the points to be included in your answer, and deciding on the order in which they should be presented, before starting to write the answer. Your answer should be written down accurately and concisely. You will lose marks by making vague statements, and you will run short of time if you use an *excessive* number of words. Make sure the words you use are essential and have meaning. If you have to describe an experiment, not only describe the apparatus but how any measurements are made. For example, if a stopwatch is used to time 20 swings of a pendulum, say just that. By all means use diagrams to illustrate your answer, but they should be drawn clearly and neatly and labelled; otherwise they are useless.

If a calculation is involved, make sure you write down the physical principle you are using, in words or in the form of an equation, before starting on the arithmetic. At the end of the calculation see that your answer is physically reasonable. For example, it would not be reasonable to give the mass of a man as 600 kg; it is more than likely that you have made an arithmetical mistake and the answer should be 60 kg. If in doubt about your answer, check your Physics and your Arithmetic. Make sure your answer is followed by the correct unit.

Most Examining Groups now print the mark allocation at the right of the question. If a question has several parts, the marks are a guide to the relative importance of the parts, and thus the time you should spend on each.

# PRACTICE IN ANSWERING EXAMINATION QUESTIONS

This section of the book is designed to give you practice in answering different types of examination questions. It is therefore arranged in three groupings: multiple choice questions, questions requiring short answers, and longer questions.

## Multiple choice questions (answers p 137)

### Type 1

Each question below has five possible answers labelled A, B, C, D, E. For each numbered question select the one which is the best answer. Within the group of questions each answer may be used once, more than once or not at all.

*Questions 1–3*
The following graphs are drawn to show the motion of four sprinters in a race. They are all drawn to the same scale.

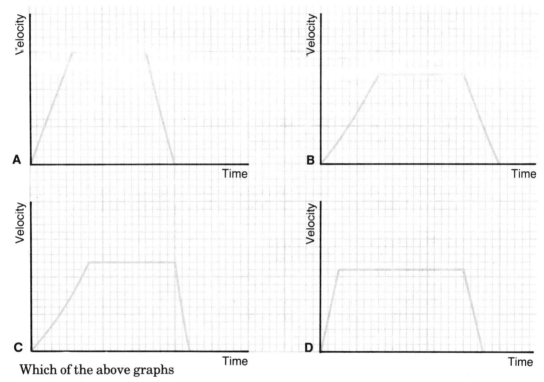

Which of the above graphs

1 shows the runner who reaches the greatest velocity?
2 shows the runner who takes the longest time?
3 shows the runner with the greatest acceleration at the start of the race?

(LEAG, Syll. B)

*Questions 4–7*
The following are five physical quantities:
  A force     C pressure      E work
  B power     D acceleration
  Which one of the above

4 involves area?
5 has the same units as weight?
6 has the same units as energy?
7 could be expressed in newton metres per second?

*Questions 8–11*
The following units are used in mechanics:
  A kilograms per cubic metre ($kg/m^3$)   D newton (N)
  B metre (m)                                E newtons per square metre ($N/m^2$)
  C joule (J)

Which unit is used to measure the following?

8 energy
9 weight
10 pressure
11 density

*Questions 12–15*
The following are types of electromagnetic radiation:
- A radio waves
- B ultraviolet radiation
- C visible light
- D infrared radiation
- E X-rays

Which type of radiation

12 passes through a thin sheet of lead?
13 causes suntan?
14 is given out from an electric fire?
15 has the longest wavelength?

*Questions 16–18*
The following are associated with atoms or molecules:
- A electrons
- B ions
- C isotopes
- D neutrons
- E protons

Which of the above

16 form cathode rays?
17 are either positively or negatively charged?
18 are the nuclei of hydrogen atoms?

## Type 2

Each of the following questions or incomplete statements is followed by four or five suggested answers labelled A, B, C, D, (E). Select the best answer in each case.

19 The perspex box in the diagram contains 4 cm depth of water. When a stone is put into the water, the surface of the water rises 2 cm further up the box. The stone is completely under water. What is the volume of the stone?

A  25 cm³    B  30 cm³
C  50 cm³    D  150 cm³

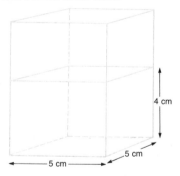

(SEG, Syll. A)

20 A ticker timer makes 50 dots every second.

The tape shown (drawn in actual size) is pulled through this timer. What speed does the tape show?

A  10 cm/s    B  50 cm/s    C  100 cm/s    D  250 cm/s    (LEAG, Syll. B)

21 The figure shows a trolley on a bench. A string attached to the trolley passes over a frictionless pulley with a scale pan attached to the end of the string.

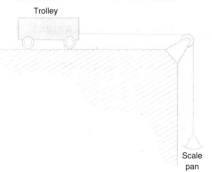

The scale pan has a weight of 0.2 N. When a weight of 0.8 N is added to the pan and the trolley is given a slight push to the right, the trolley continues moving at constant speed.

The frictional force opposing the motion of the trolley is

   A 0.2 N      B 0.6 N      C 0.8 N      D 1.0 N      (NEA, Syll. A)

*Questions 22 and 23*

You are the officer in charge of a spaceship travelling from our Galaxy to the Andromeda Galaxy. You are in deep space, the gravitational force due to surrounding galaxies is zero.

22 If your drive motors are off then you must be

   A stationary
   B decreasing in speed
   C increasing in speed
   D maintaining your present speed

23 Your spaceship has a mass of $10^6$ kg. If the rockets produce a thrust of $10^5$ N what is your acceleration?

   A 0.1 m/s$^2$      B 1 m/s$^2$      C 10 m/s$^2$      D 11 m/s$^2$      (LEAG, Syll. A)

24 A body accelerates from rest at 4 m/s$^2$ for 5 s. What is its average speed?

   A 0.8 m/s      B 1.2 m/s      C 9 m/s      D 10 m/s      E 11 m/s

25 The diagram shows an aeroplane. Which lettered arrow shows the direction of the gravitational force acting on the aeroplane?

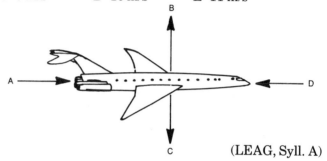

(LEAG, Syll. A)

26 The mass of a moving body multiplied by its velocity is measuring the body's

   A inertia      C acceleration      E energy
   B weight      D momentum

27 In which of the following situations would a lorry have the greatest momentum?

   A stationary with no load      C moving at 10 m/s with no load
   B stationary with a load      D moving at 10 m/s with a load

(LEAG, Syll. B)

28 To drag a 25 kg packing case across a floor needs a horizontal force of 100 N. How much work is done in moving the case 2 metres?

   A 50 J      B 100 J      C 200 J      D 500 J      (SEG, Syll. A)

29 Which one of the following units could be used to measure the rate at which a machine is doing work?

   A joule      C newton metre      E watt
   B newton      D newton per second

30 Which list of energy sources, A to D, contains only non-renewable sources?

   A wind, wave, nuclear      C coal, oil, geothermal
   B hydroelectric, gas, oil      D gas, oil, nuclear      (NEA, Syll. A)

31 An object is dropped from a tower 50 m high. When it is 10 m from the ground what is the ratio of the body's potential energy to its kinetic energy?

   A 1:5      B 1:4      C 1:2      D 4:1      E 5:1

32 A large body of mass 2 kg falls from rest at a point 5 metres above the level of the ground at a speed of 8 m/s. If the body does not rebound, what quantity of energy is converted into heat and sound on impact?

   A 20 J      B 64 J      C 100 J      D 128 J      (NEA, Syll. A)

*Questions 33 and 34*

John's weight is 600 N. He runs at a constant speed up a flight of stairs of total height 3 metres in 5 seconds.

33 The work John does is

   A 9000 J      B 1800 J      C 360 J      D 40 J

34 John's power is
   A  9000 W     B  1800 W     C  360 W     D  40 W               (LEAG, Syll. A)

35 Walking to the top of a steep hill usually seems to be easier if a zig-zag path is used instead of climbing by the steepest route. Why is this?
   A  Less energy is used
   B  Less friction has to be overcome
   C  Less power is needed
   D  Less time is wasted                                          (SEG, Syll. A)

36 An aircraft in flight is subject to four forces.

   In order for the aircraft to accelerate forward in level flight
   A  drag must be greater than thrust
   B  lift must be greater than weight
   C  thrust must be greater than drag
   D  thrust must be greater than weight                           (SEG, Syll. A)

37 A girl weighing 600 newtons sits 6 metres away from the pivot of a see-saw, as shown above. What force $F$, 9 metres from the pivot, is needed to balance the see-saw?
   A  300 newtons     C  450 newtons     E  900 newtons
   B  400 newtons     D  600 newtons

38 The diagram shows four designs for table lamps. In each case the position of the centre of gravity is shown by point G. Which of the designs A, B, C, or D is the most stable?
                                                                   (SEG, Syll. A)

39 A machine is a device which normally:
   A  increases power     C  saves energy     E  creates forces
   B  saves time          D  magnifies forces

40 An electric crane does 1500 J of useful work when the energy input is 2000 J. The efficiency of the crane is
   A  0.25 (25%)     C  0.75 (75%)
   B  0.5 (50%)      D  1 (100%)                                   (LEAG, Syll. A)

41 A solid ball is taken from the earth to the moon. On the moon, the ball will have a very different
   A  density     B  mass     C  volume     D  weight              (SEG, Syll. A)

42 A solid cube of material of density 0.5 g/cm³ has sides of length 2 cm.

The mass of the cube is

A 1 g    B 2 g    C 4 g    D 16 g

(LEAG, Syll. A)

43 A stone has a mass of 480 g and a volume of 160 cm³. Its density is:

A 76 800 g/cm³    C 320 g/cm³    E 0.3 g/cm³
B 540 g/cm³       D 3 g/cm³

44 A car's tyres are inflated so that an area of 50 cm² of each is in contact with the ground. If the mass of the car is 1600 kg and is evenly distributed between the four tyres, what is the pressure exerted on the ground by each one? $g = 10$ N/kg

A 80 N/cm²    D 320 N/cm²
B 160 N/cm²   E 500 N/cm²
C 200 N/cm²

45 Lemonade enters the mouth through a drinking straw when the drinker breathes in. This is because

A it is drawn up by the pull of the air in the straw.
B the straw acts as a capillary.
C it is sucked up by the vacuum in the lungs.
D it is pushed up by the atmospheric pressure.

(NEA, Syll. A)

46 The diagram shows part of a hydraulic brake.

When the brake pedal is pressed the pressure of the brake fluid

A acts equally in all directions.
B is greatest to the right.
C is greatest to the left.
D is greatest upwards.    (LEAG, Syll. A)

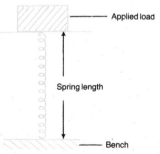

47 Atmospheric pressure decreases with increasing height above the earth's surface. Which of the following instruments could be used to show this decrease?

A barometer     C newtonmeter
B manometer     D thermometer

(SEG, Syll. A)

48 The diagram shows a spring being tested with a load.

Which of the graphs of spring length against load is correct?

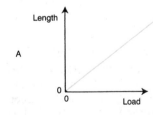

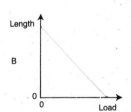

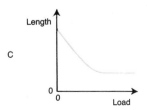

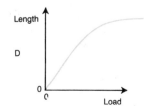

(SEG, Syll. A)

49 A force of 2 newtons stretches a spring by 5.0 cm. Two such springs are put side by side as shown and the same load of 2 newtons is applied.

By how much is the pair of springs stretched?

A  2.5 cm    B  4.0 cm    C  5.0 cm    D  10.0 cm

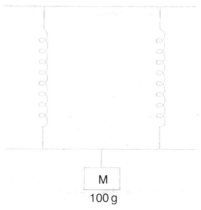

(SEG, Syll. A)

*Questions 50–51*
Two springs are such that, when either is loaded with a mass of 200 g it stretches 4 cm. The masses of the springs may be ignored.

50 The springs are suspended in parallel and a load of 100 g is attached. What is the most likely stretch of each spring?

A  0.5 cm    B  1 cm    C  2 cm    D  4 cm    E  8 cm

51 The springs are suspended in series and a load of 200 g is attached. What is the total stretch of both springs together?

A  0.5 cm    B  1 cm    C  2 cm    D  4 cm    E  8 cm

52 Which of the following describes particles in a solid?

A  close together and stationary
B  close together and vibrating
C  close together and moving around at random
D  far apart and stationary
E  far apart and moving around at random

53 'Brownian motion' is the name for the movement of

A  a gas diffusing into a vacuum
B  electrons in an atom
C  ions in electrolysis
D  molecules in a heated metal bar
E  smoke particles in air

54 The diagram shows a simplified view of sections of a motorway bridge. Sections are supported on rollers and have a small gap between them.

The purpose of the rollers and gap arrangement is

A  to allow rain water to drain away from the road surface.
B  to allow for expansion of the bridge sections when the temperature changes.
C  to allow for vibrations of the bridge due to the effects of wind.
D  to allow for force changes in the bridge as heavy lorries drive across.

(LEAG, Syll. A)

55 The graph shows how three metal strips X, Y and Z vary in length as they are heated.

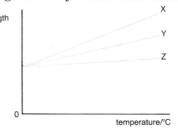

Bimetallic strips are made from pairs of these metals. Which strip will bend upwards when it is heated?

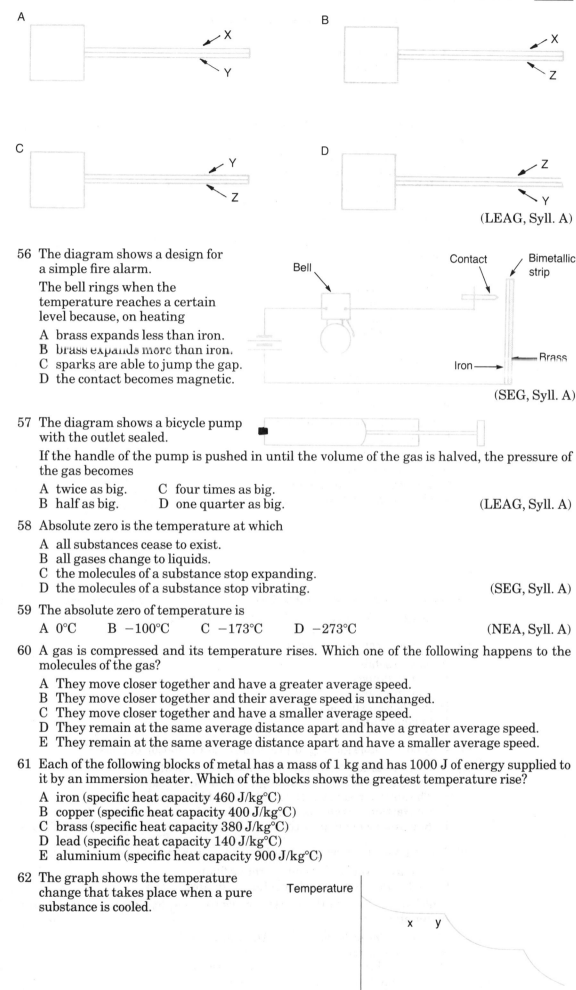

(LEAG, Syll. A)

56 The diagram shows a design for a simple fire alarm.

The bell rings when the temperature reaches a certain level because, on heating

A brass expands less than iron.
B brass expands more than iron.
C sparks are able to jump the gap.
D the contact becomes magnetic.

(SEG, Syll. A)

57 The diagram shows a bicycle pump with the outlet sealed.

If the handle of the pump is pushed in until the volume of the gas is halved, the pressure of the gas becomes

A twice as big.   C four times as big.
B half as big.    D one quarter as big.

(LEAG, Syll. A)

58 Absolute zero is the temperature at which
A all substances cease to exist.
B all gases change to liquids.
C the molecules of a substance stop expanding.
D the molecules of a substance stop vibrating.

(SEG, Syll. A)

59 The absolute zero of temperature is
A 0°C    B −100°C    C −173°C    D −273°C

(NEA, Syll. A)

60 A gas is compressed and its temperature rises. Which one of the following happens to the molecules of the gas?

A They move closer together and have a greater average speed.
B They move closer together and their average speed is unchanged.
C They move closer together and have a smaller average speed.
D They remain at the same average distance apart and have a greater average speed.
E They remain at the same average distance apart and have a smaller average speed.

61 Each of the following blocks of metal has a mass of 1 kg and has 1000 J of energy supplied to it by an immersion heater. Which of the blocks shows the greatest temperature rise?

A iron (specific heat capacity 460 J/kg°C)
B copper (specific heat capacity 400 J/kg°C)
C brass (specific heat capacity 380 J/kg°C)
D lead (specific heat capacity 140 J/kg°C)
E aluminium (specific heat capacity 900 J/kg°C)

62 The graph shows the temperature change that takes place when a pure substance is cooled.

What change of state does the section $X-Y$ of the graph represent?

A a solid to a gas
B a liquid to a solid
C a solid to a liquid
D a gas to a liquid
E a liquid to a gas

63 Two liquids are spilt on the hand. One is alcohol, the other is water, and both are at the same temperature. Why does the alcohol feel colder?

A It has a higher boiling point than water.
B It is a worse conductor of heat than water.
C It has a higher specific latent heat of vaporization than water.
D It evaporates more readily than water.
E It cools on evaporating.

64 When more molecules of a liquid return to it than escape from it, what is said to be happening?

A It is condensing.
B It is conducting.
C It is diffusing.
D It is evaporating.
E It is radiating.

65 The figure shows water boiling at the top of a glass test tube while an ice cube remains unmelted at the bottom.

From this demonstration it may be deduced that

A water transfers heat by conduction.
B glass is a good conductor of heat.
C water is a poor conductor of heat.
D glass transfers heat by convection.

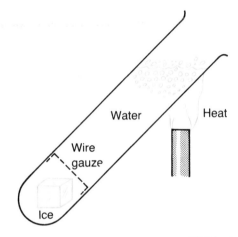

(NEA, Syll. A)

66 Which colour is most suitable for use in a solar panel which absorbs the sun's radiation?
   A black     B silver     C white     D yellow     (SEG, Syll. A)

67 Where does convection occur?

A only in solids
B only in liquids
C only in gases
D in solids and liquids
E in liquids and gases

68 Why does double glazing improve the heat insulation of houses?

A Glass of double thickness does not conduct heat.
B Radiation will not pass through two sheets of glass.
C The air trapped between the glass is a bad conductor of heat.
D Convection currents between the glass sheets are restricted.
E Radiation will not pass through the gap between the two sheets of glass.

69 Two ways of reducing heat loss from houses are as follows

(i) fill cavity walls with polyurethane foam.
(ii) put shiny aluminium foil behind radiators.

Which of the following correctly shows the heat loss processes which are reduced by these materials?

|   | Polyurethane foam | Aluminium foil |
|---|---|---|
| A | conduction | radiation |
| B | conduction | conduction |
| C | radiation | radiation |
| D | radiation | conduction |

(LEAG, Syll. A)

*Questions 70 and 71*

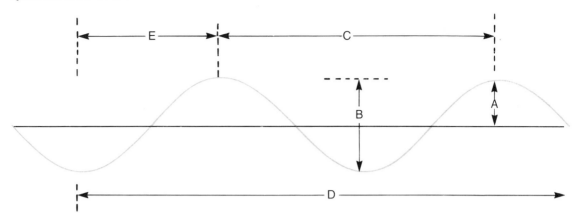

The diagram shows a wave. Which letter shows

70 the wavelength?

71 the amplitude?

72 If a cork is placed on gentle water waves of constant frequency how will the cork move?

    A side to side at irregular intervals    C up and down at irregular intervals
    B side to side at regular intervals    D up and down at regular intervals

(SEG, Syll. A)

73 Which one of the following radiations is not transverse?

    A radio    B sound    C infrared    D light    E gamma-rays

74 Which one of the following statements about the image produced in a pinhole camera is correct?

    A The image is bigger if the object is further away.
    B The image is smaller if the screen is nearer the pinhole.
    C The image is brighter if the object is further away.
    D The image is sharper if the pinhole is made bigger.
    E The image is bigger if the object is brighter.

75 When you look at yourself in a mirror you see an image of yourself. The image is

    A on the surface of the mirror    D caused by rays behind the mirror
    B a real image behind the mirror    E a virtual image behind the mirror
    C an inverted virtual image

76 A photographer wishes to take pictures without being noticed. He attaches two plane mirrors to his camera.

Which arrangement of mirrors will allow the photographer to take pictures of someone behind the camera? (LEAG, Syll. A)

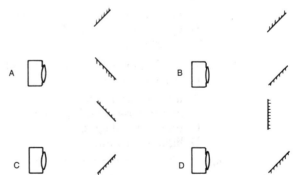

77 A model car, driven by an electric motor, is used to make a model of refraction.

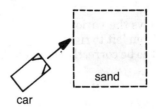

The car is placed on a laboratory bench and runs towards an area of sand. The car travels slowly through the sand. Which diagram (overleaf) shows the most likely path of the car?

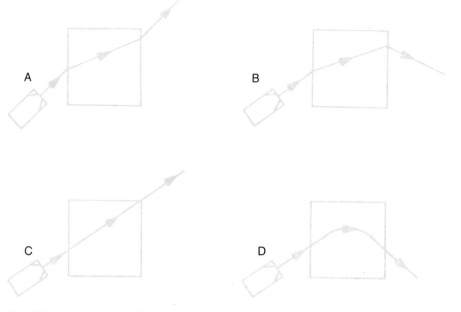

(LEAG, Syll. A)

78 Which one of the following changes occurs when a light wave passes from air into glass?

    A Its wavelength increases.    D Its frequency decreases.
    B Its speed decreases.    E Its frequency increases.
    C Its speed increases.

79 Light can travel from one end of an optical fibre to another, even if the fibre follows a curved path. This is due to

    A dispersion of the light.    C the use of plane mirrors.
    B refraction of the light.    D total internal reflection.    (SEG, Syll. A)

80 Which of the following describes the image seen through a magnifying glass?

    A It is real and upright.    D It is virtual and inverted.
    B It is real and inverted.    E It is none of these.
    C It is virtual and upright.

81 Which of the following describes the image formed on the retina of the human eye?

    A It is magnified and erect.    D It is diminished and erect.
    B It is diminished and inverted.    E It is magnified and inverted.
    C It is life-size and virtual.

82 Light waves spread out when they pass through a very narrow gap. What is this effect called?

    A refraction    D diffraction
    B reflection    E dispersion
    C interference

83 An explosion emits light, infrared and sound waves. Which waves will be received first by a person who is some distance away?

    A infrared and sound
    B all three received together
    C infrared and light
    D light and sound    (LEAG, Syll. A)

84

| Radio | 1 | Visible | Ultra violet | 2 | 3 |
|---|---|---|---|---|---|

The chart shows the various parts of the electro-magnetic spectrum, in order of decreasing wavelength from left to right.

For the chart to be correctly completed, what should be inserted in regions 1, 2, and 3 on the chart?

|   | 1 | 2 | 3 |
|---|---|---|---|
| A | gamma-rays | infrared | X-rays |
| B | infrared | X-rays | gamma-rays |
| C | X-rays | gamma-rays | infrared |
| D | infrared | gamma-rays | X-rays |

(LEAG, Syll. A)

85 Which of the following examples of electromagnetic radiation has the shortest wavelength?
   A radio waves      D visible light
   B infrared rays    E X-rays
   C ultraviolet rays

86 Which of the following does NOT apply to sound waves?
   A They transmit energy.
   B They result from vibrations.
   C They are propagated by a series of compressions and rarefactions.
   D They travel fastest in a vacuum.
   E They can be diffracted.

87 A loudspeaker consists of a cone which vibrates to and fro.
   If the cone is made to vibrate with a bigger amplitude, the sound will become
   A higher pitched
   B lower pitched
   C louder
   D quieter

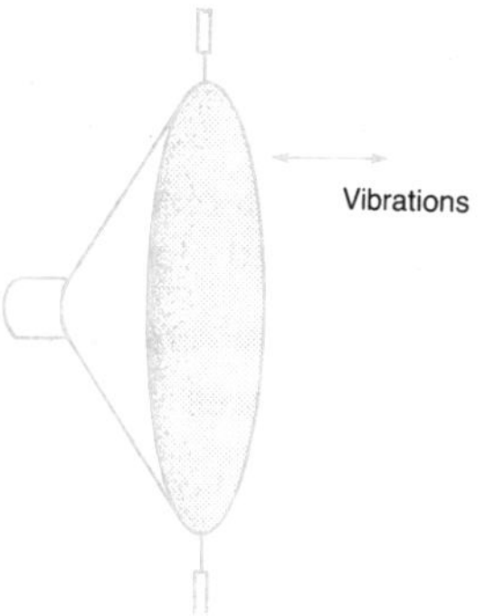

   (LEAG, Syll. A)

88 When the foghorn on the lighthouse in the figure sounded a short blast, the echo was heard at the lighthouse 5 seconds later. The speed of sound was
   A 165 m/s    B 330 m/s    C 825 m/s    D 4125 m/s    (NEA, Syll. A)

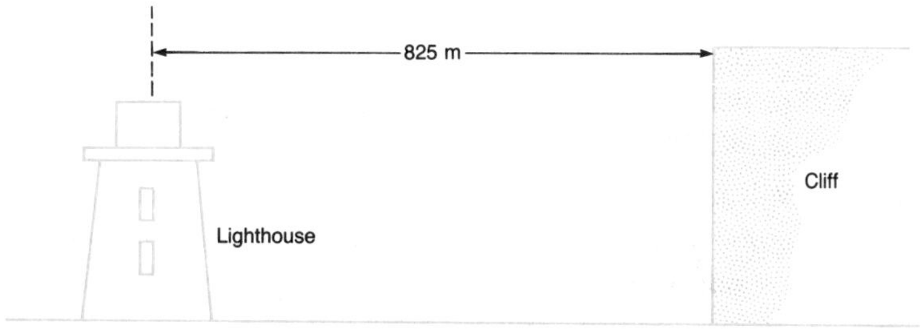

89 How much time is there between a person firing a starting pistol and being able to hear the echo from a wall 150 m away? The speed of sound is 300 m/s.
   A 0.5 s    B 1.0 s    C 1.5 s    D 3.0 s    (SEG, Syll. A)

90 When the pitch of a note is raised, which property of the note is increased?
   A frequency    B wavelength    C amplitude    D speed    (NEA, Syll. A)

91 Two magnets are hung as shown in the figure and are seen to repel each other.
   This shows that X and Y must
   A both be north poles.
   B be unlike poles.
   C both be south poles.
   D both be south poles or both be north poles.
   (NEA, Syll. A)

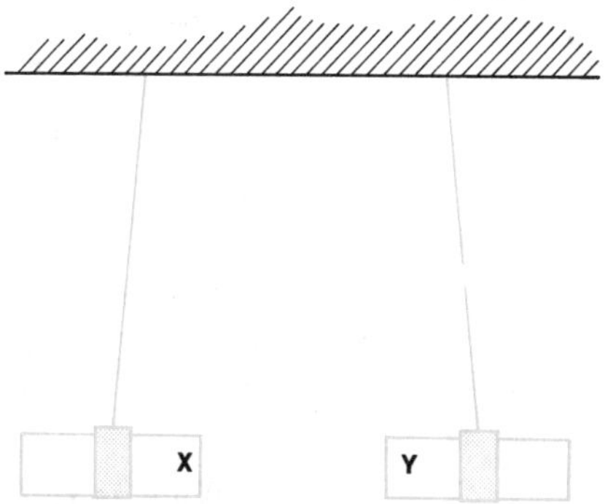

92 The diagrams show different arrangements of two magnets. All the magnets are equally strong. Which pair of magnets will pull together?

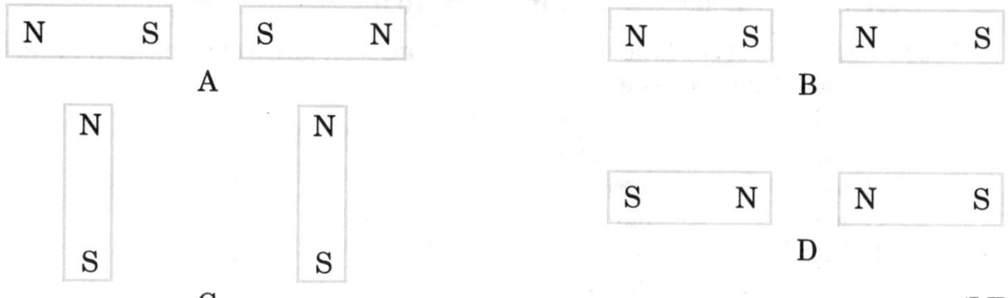

   (LEAG, Syll. A)

93 Which of the following have to be brought together to produce the magnetic field pattern shown in the diagram?

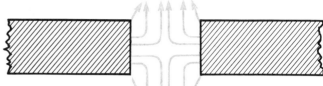

A  N-pole and S-pole    D  S-pole and N-pole
B  N-pole and N-pole    E  S-pole and unmagnetized bar
C  S-pole and S-pole

94 Which of the following could be used as an electrical insulator?
A  Carbon    B  Copper    C  Lead    D  Plastic          (SEG, Syll. A)

95 When Darren takes his jumper off, it becomes negatively charged as it rubs against his shirt. This is because the jumper has
A  lost electrons.    C  lost protons.
B  gained electrons.    D  gained protons.          (LEAG, Syll. A)

96 Electric charge is measured in
A  amperes    B  coulombs    C  ohms    D  volts          (NEA, Syll. A)

97 When an electric current flows in a metal, charged particles move. What are these particles called?
A  atoms    B  electrons    C  ions    D  protons          (SEG, Syll. A)

98 Which of the following is equivalent to 1 ampere?
A  1 coulomb per second    D  1 joule per second
B  1 coulomb per watt    E  1 watt per second
C  1 joule per coulomb

99 One volt is

A  $\dfrac{\text{one ampere}}{\text{one ohm}}$    B  $\dfrac{\text{one ampere}}{\text{one watt}}$    C  $\dfrac{\text{one ohm}}{\text{one ampere}}$    D  $\dfrac{\text{one watt}}{\text{one ampere}}$

(SEG, Syll. A)

100 The current flowing through a filament lamp varies with the potential difference across it as shown in the graph.

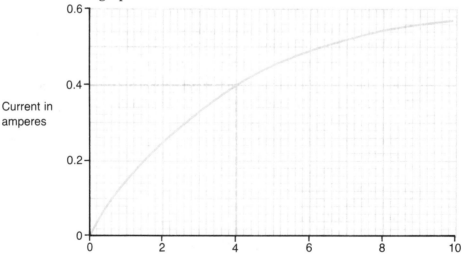

What is the resistance of the lamp, in ohms, when the potential difference across it is 4 volts?
A  0.1    B  0.4    C  1.0    D  10.0          (SEG, Syll. A)

101 Which one of the following is true for the resistance of a voltmeter and the way in which it is normally connected in a circuit?

| | resistance | connection |
|---|---|---|
| A | zero | parallel |
| B | low | series |
| C | low | parallel |
| D | high | series |
| E | high | parallel |

102 The circuit diagram shows three resistors in series and four voltmeters; the voltmeters are measuring potential differences between various points in the circuit.

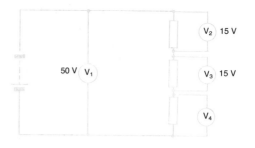

The reading on $V_1$ is 50 V; that on $V_2$ is 15 V and that on $V_3$ is also 15 V. The reading on $V_4$ must be

A  15 V    B  20 V    C  30 V    D  80 V

(LEAG, Syll. A)

103 The diagrams show two pairs of resistors connected together.

Which of the following correctly describes the connections in each case?

```
    R₁ and R₂      R₃ and R₄
A   series         parallel
B   series         series
C   parallel       series
D   parallel       parallel
```

(LEAG, Syll. A)

104 What is the total resistance between **X** and **Y** in the circuit shown?

A  50 ohms    B  100 ohms    C  200 ohms    D  10 000 ohms

(SEG, Syll. A)

*Questions 105 and 106*
The diagram shows a circuit containing a lamp L, a voltmeter and ammeter.

The voltmeter reading is 3 V and the ammeter reading is 0.5 A.

105 What is the resistance of the lamp?

A  1.5 Ω    B  2.5 Ω    C  4.5 Ω    D  6.0 Ω

106 What is the power of the lamp?

A  1.5 W    B  2.5 W    C  4.5 W    D  6.0 W

(LEAG, Syll. A)

107 What is the most probable value of the current flowing through a 12 V 24 W car headlamp when operating on a 6 V supply?

A  0.2 A    B  1 A    C  1.2 A    D  2 A    E  4 A

108 An electricity board charges 6 p per kilowatt-hour (kWh) of electrical energy supplied. What is the total cost of operating 4 light bulbs, each rates at 100 W, for 5 hours?

A  3 p    B  6 p    C  12 p    D  120 p

(SEG, Syll. A)

109 Electricity costs 6 p per kilowatt-hour. If you run a 3 kW fire for five hours in an evening, how much will it cost?

A  2.5 p    B  3.6 p    C  10 p    D  90 p

(LEAG, Syll. A)

110 A light on an upstairs landing is controlled by a downstairs and an upstairs 2-way switch as shown.

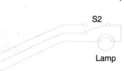

Which line A, B, C or D in the following table gives the correct positions of the switches and shows if the lamp is on or off?

|   | Position of $S_1$ | Position of $S_2$ | Lamp on or off |
|---|---|---|---|
| A | down | down | off |
| B | down | up | on |
| C | up | down | off |
| D | up | up | off |

(SEG, Syll. A)

111 Fuses are available which melt when the current through them exceeds 2, 5, 10, 15 or 30 A. Which one of these would be the most suitable for a circuit in which an electric fire rated at 2.5 kW is to be connected, if the supply voltage is 240 V?

A 2 A fuse
B 5 A fuse
C 10 A fuse
D 15 A fuse
E 30 A fuse

112 The earth wire to an electric iron should be connected to

A one end of the element.   C the plastic handle.
B the metal casing.   D the thermostat.

(SEG, Syll. A)

113 Small compasses P, Q, R and S are placed around a rod-shaped electromagnet to show the direction of the magnetic field at various places.

If the current through the electromagnetic is reversed how will the needles react?

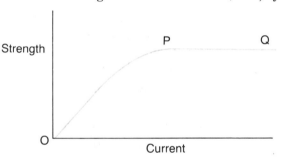

A P, Q, R and S will all reverse direction
B P, Q, R and S will all remain in the same direction
C P and S will reverse, Q and R would remain unchanged
D Q and R will reverse, P and S would remain unchanged

(SEG, Syll. A)

114 The diagram shows the strength of an electromagnet with a fixed number of turns of wire as a function of the steady current passing through the turns.

Why can the strength of the electromagnet not be increased above the value of the line PQ?

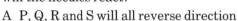

A The turns of wire get too hot.
B The current through the turns cannot be increased further.
C The small magnets (domain) within the core have all be aligned by point $P$.
D The core gets too hot.
E The core is made of a material which is unsuitable for use in an electromagnet.

115 Which of the following changes would not affect the direction of the force on a wire carrying a current in a magnetic field?

A The direction of the magnetic field is reversed.
B The magnetic field is removed.
C The direction of the current is reversed.
D The current is switched off.
E Both the direction of the magnetic field and the direction of the current are reversed.

116 The diagram shows a current flowing through a wire placed in a magnetic field. The strength of the force acting on the wire is $F$.

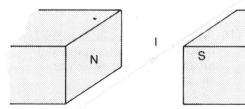

The current flowing through the wire is now halved and the strength of the magnetic field is doubled. Which one of the following statements is true about the force now acting on the wire?

A It has strength $4F$ and is vertical.
B It has strength $4F$ and is horizontal.
C It has strength $2F$ and is horizontal.
D It has strength $F$ and is vertical.
E It has strength $F$ and is horizontal.

117 If a bar magnet is pushed into a solenoid, an electromotive force (e.m.f.) will be induced across the ends of the solenoid. This e.m.f. can be made larger by

A using a bar of soft iron instead of a magnet.
B using a solenoid made of low resistance wire.
C moving the magnet more quickly.
D connecting a voltmeter across the solenoid. (SEG, Syll. A)

118 For which one of the following is an alternating current essential in its operation?

A an electromagnet     C a galvanometer     E an electric fire
B a transformer        D an electric lamp

119 Which of the following diagrams shows a transformer that can change 240 V a.c. into 12 V a.c? (LEAG, Syll. A)

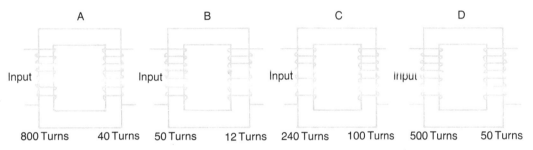

120 A transformer is used to change mains 240 volts a.c. to 12 volts a.c. for the heater of a fish tank. If the primary coil of the transformer has 2400 turns, how many turns must there be on the secondary coil?

A 120     B 1200     C 2400     D 4800 (SEG, Syll. A)

121 A step down transformer has a turns ratio of 6:1. A 240 V alternating current supply at a frequency of 50 Hz is connected to the primary coil. What is the frequency of the alternating current output?

A 0.2 Hz     B 5 Hz     C 10 Hz     D 50 Hz     E 500 Hz

122 The diagram shows the main parts of a power station. Some energy is wasted at each stage and the efficiency of each part is shown.

How many joules of electrical energy are produced?

A 360 J     B 400 J     C 800 J     D 900 J (SEG, Syll. A)

123 What is emitted from an incandescent filament during thermionic emission?

A electrons     B heat     C infrared radiation     D ions     E light

124 An alternating potential difference (p.d.) is applied to the Y-plates of a cathode ray oscilloscope with the time-base switched on.

The appearance of the screen is shown in Figure 1. In order to change the appearance of the trace to that shown in Figure 2 it is necessary to

A decrease the speed of the time-base.
B decrease the frequency of the applied p.d.
C increase the speed of the time-base.
D decrease the applied p.d. (NEA, Syll. A)

125 A thermistor is a semi-conductor device

A which converts electrical energy into light.
B whose resistance depends on the amount of light incident on it.
C whose resistance depends on its temperature.
D which converts light into electrical energy. (NEA, Syll. A)

126 Shabnam wants to construct a simple transistor switch that will operate a buzzer when it becomes light in the morning. Which of the following should she use to detect a change in light intensity?

    A filament lamp      C light emitting diode
    B light dependent resistor      D semiconductor diode      (LEAG, Syll. A)

*Questions 127 and 128*
The diagram shows a circuit for a transistor switch.

127 Which of the lettered symbols on the diagram represents the transistor?

128 In this circuit the transistor is switched on by a change in

    A temperature      B light intensity      C pressure      D moisture level
    (LEAG, Syll. A)

129 In which of the following circuits will the LED light?

(LEAG, Syll. A)

130 This circuit could be used as a fire alarm. The four components labelled A to D are named below.

    A thermistor
    B variable resistor
    C lamp
    D transistor

Which component is the sensor which detects the heat from the fire?

(LEAG, Syll. A)

131 What is a particle with a mass of 1 atomic unit and a charge of +1 called?

    A an electron      C a positron      E an alpha particle
    B a gamma particle      D a proton

132 Which one of the following quantities in an atom represents the mass number of the atom?

    A the number of protons plus the number of neutrons      D the number of protons
    B the number of protons plus the number of electrons      E the number of neutrons
    C the number of electrons

133 The proton (atomic) number of an atom is equal to

    A the mass of the atom.
    B the number of neutrons in the nucleus.
    C the number of electrons in the atom.
    D the total charge on the atom.      (SEG, Syll. A)

134 A spent fuel rod containing radioactive material from a nuclear reactor has an activity of 40 000 Bq. It must be stored until the activity is 20 000 Bq. The half life of the radioactive material is 1 month. For how long must the fuel rod be stored? (1 Bq = 1 count per second.)

    A 2 weeks      B 1 month      C 2 months      D 4 months      (LEAG, Syll. A)

135 What does an atom become when it loses an electron?

    A a negative ion      C a positive ion      E a proton
    B a neutron      D an alpha particle

136 An element $^{238}_{92}$X decays to $^{234}_{90}$Y by radioactive emission. The emission is
A an alpha particle   C a neutron
B a beta particle   D a proton   (LEAG, Syll. A)

137 A radiographer must work near a radioactive source which emits rays. Which of the following will provide the greatest protection?
A a thin aluminium screen   C a thin lead screen
B a thick aluminium screen   D a thick lead screen   (LEAG, Syll. A)

138 When radioactive substances are being stored, they are kept inside closed boxes. These boxes are lined with lead because lead
A greatly reduces the amount of radiation getting out of the box.
B makes the box heavy so that it cannot easily be stolen.
C prevents magnetic fields affecting the strength of the substance.
D stops light getting to the radioactive substance.   (SEG, Syll. A)

## Answers to multiple choice questions

| | | | | | | | | | |
|---|---|---|---|---|---|---|---|---|---|
| 1 A | 2 B | 3 D | 4 C | 5 A | 6 E | 7 B | 8 C | 9 D | 10 E |
| 11 A | 12 E | 13 B | 14 D | 15 A | 16 A | 17 B | 18 E | 19 C | 20 C |
| 21 D | 22 D | 23 A | 24 D | 25 C | 26 D | 27 D | 28 C | 29 E | 30 D |
| 31 B | 32 B | 33 B | 34 C | 35 C | 36 C | 37 B | 38 C | 39 D | 40 C |
| 41 D | 42 C | 43 D | 44 A | 45 D | 46 A | 47 A | 48 C | 49 A | 50 B |
| 51 E | 52 B | 53 E | 54 B | 55 D | 56 B | 57 A | 58 D | 59 D | 60 A |
| 61 D | 62 D | 63 D | 64 A | 65 C | 66 A | 67 E | 68 C | 69 A | 70 C |
| 71 A | 72 D | 73 B | 74 B | 75 E | 76 C | 77 A | 78 B | 79 D | 80 C |
| 81 B | 82 D | 83 C | 84 B | 85 E | 86 D | 87 C | 88 B | 89 B | 90 A |
| 91 D | 92 B | 93 B | 94 D | 95 B | 96 B | 97 B | 98 A | 99 D | 100 D |
| 101 E | 102 B | 103 C | 104 A | 105 D | 106 A | 107 D | 108 C | 109 D | 110 C |
| 111 D | 112 B | 113 A | 114 C | 115 E | 116 D | 117 C | 118 B | 119 A | 120 A |
| 121 D | 122 A | 123 A | 124 A | 125 C | 126 B | 127 D | 128 A | 129 A | 130 A |
| 131 D | 132 A | 133 C | 134 B | 135 C | 136 A | 137 D | 138 A | | |

## Questions requiring short or structured answers (answers p 180)

Questions of the short answer or structured type are used by most of the Examining Groups. The answer can be a single word, or number, a phrase, or a few sentences. A space is usually left for the student's answer. This is often an indication of the length of answer required. Some groups allocate marks, which can also indicate the length of answer required and the amount of time you should spend on each part of the question.

### Unit 2
1 An experiment is carried out to study the motion of a trolley along a runway using the apparatus in the diagram. The runway is friction compensated.

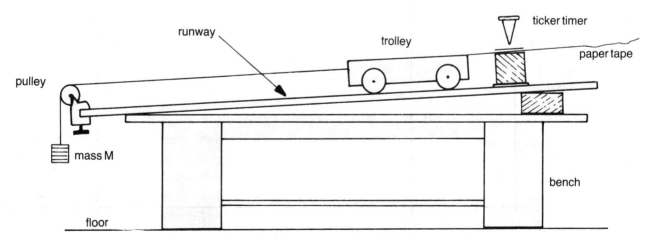

The trolley is released from rest. It is pulled along the runway as the mass, M, falls. After M hits the floor the trolley continues to the end of the runway.

The tape from the experiment is cut into 'five-tick' lengths and used to make a 'tape chart' of the trolley's motion. The chart is shown below. It is full size.

**138** *Short answer questions*

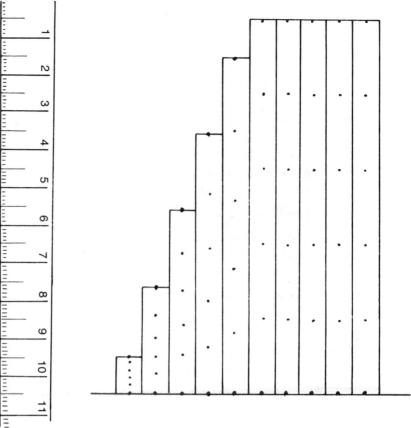

scale marked in centimetres

(a) Describe the motion of the trolley along the runway. (3 marks)
(b) (i) Measure the 'tape chart' to find the longest 'five-tick' length.
(ii) The ticker timer makes 50 dots per second. Calculate the **greatest** speed reached by the trolley. (5 marks)
(c) Why does the trolley **not** speed up during the final part of the journey? (2 marks)
(d) Underline the correct answer in each of the three sentences below:
The mass, M, attached to the trolley is doubled and the experiment is repeated.
(i) The acceleration of the trolley will (increase/stay the same/decrease).
(ii) The time taken for the mass, M, to reach the ground will (increase/stay the same/decrease).
(iii) The final speed of the trolley will (increase/stay the same/decrease). (3 marks)
(Total: 13 marks)
(LEAG, Syll. B, Paper 2)

2 The diagram below is a graph of the journey made by a train. Its speed is measured in metres per second (m/s) and time is measured in second (s).

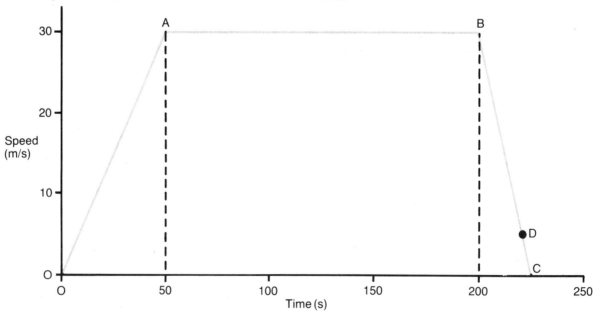

(a) What is the greatest speed reached by the train?
(b) How is the train moving on the section represented by the line $AB$?
(c) How is the train moving in the section represented by the line $BC$?
(d) Which of the points $O, A, B$ or $C$ represents the stage at which the brakes are first applied?

(e) The line BC is steeper than the line OA. What does this tell you about the rates at which the train speeds up and slows down?
(f) Calculate how far the train travelled between the stages in its journey represented by the points O and A.

3 The velocity-time graph shows part of the motion of two cars A and B.

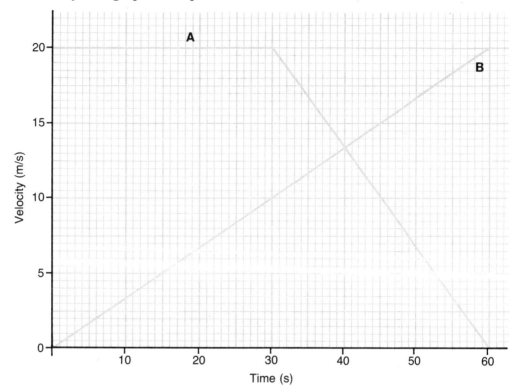

(a) Describe the motion of car A between 0 and 60 seconds. (2 marks)
(b) Write down the time when both cars are travelling at the same speed. (1 mark)
(c) Calculate
(i) the difference in the speeds of the cars at 22 s,
(ii) the acceleration of car B,
(iii) the distance car A has travelled between 30 s and 60 s. (5 marks)
(d) Car A has travelled a greater distance than car B in 60 s. How can you tell this from the graph? (1 mark)
(Total: 9 marks)
(WJEC, Paper 2A)

4 An object is released from a helicopter which is hovering (stationary) 180 m above the ground. Ignore the effect of air resistance.
(a) Calculate how long it takes the object to reach the ground?
The helicopter now starts to move horizontally at a speed of 40 m/s, and drops a second object from a height of 180 m.
(b) How long does it take the second object to reach the ground?
(c) Calculate the horizontal distance travelled by the second object between leaving the helicopter and reaching the ground.

## Unit 3

5 An oil tanker of mass 10 000 000 kg, when its engine stops, decelerates at 0.2 m/s$^2$.
(a) Calculate the size of the decelerating force on the tanker.
(b) Explain briefly how large ships are generally brought to rest. (2 marks)
(WJEC, Paper 1)

6 A hot air balloon is being used to advertise a new brand of washing powder. The figure shows a stationary balloon and basket above the ground.
(a) Draw on the diagram arrows to show any vertical forces acting on the balloon and basket. Label the arrows with the names of the forces. (2 marks)
(b) The balloon is stationary. What can you say about the upward and downward forces? (1 mark)

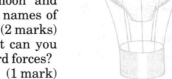

(c) The sketch-graph shows how the velocity changes with time as the balloon moves upward.

Mark on the curve, using the letters given below, the points at which the balloon
(i) has the greatest velocity (V),
(ii) has gone furthest (F),
(iii) has no acceleration (N). (3 marks)
(Total: 6 marks)
(NEA, Syll. B, Level P)

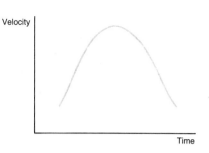

7 A free-fall parachutist jumps from an aircraft on a calm day. The figure shows a graph of his speed plotted against his time of fall.

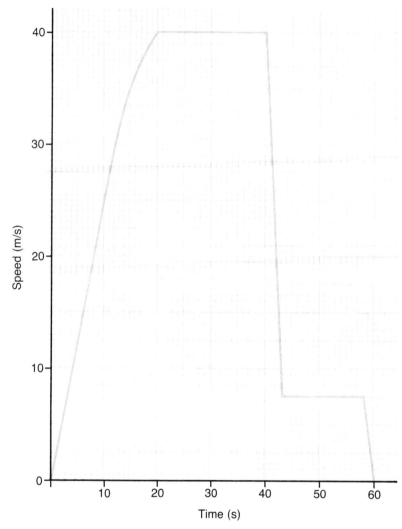

(a) (i) Name the force pulling the parachutist to the ground.
(ii) Name another force acting on the parachutist.
(iii) What happens to this other force as his speed increases? (3 marks)
(b) Between 20 s and 40 s the speed is steady.
(i) How can you tell from the graph that the speed is steady?
(ii) Suggest why the speed is steady.
(iii) Calculate how far the parachutist falls between the time of 20 s and the time of 40 s.
(8 marks)
(c) (i) What is happening to his speed between 40 s and 43 s?
(ii) Explain why the speed is changing between 40 s and 43 s.
(iii) At what speed does the parachutist touch the ground? (7 marks)
(Total: 18 marks)
(NEA, Syll. A, Level P)

8 (a) A book rests on a table. The forces acting on the book are shown in the diagram.

$R$ is the **upward** push of the **table** on the book.

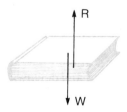

(i) W is the _____ pull of the _____ on the book.
(Use single words to fill in the blank spaces.) (2 marks)
(ii) W is usually called the _____ of the book. (1 mark)
(iii) The size of the force W is 5 N. What is the size of the force R? (1 mark)
(iv) What is the mass of the book? (2 marks)
(b) The book is now lifted off the table and allowed to fall. The diagram below shows the book as it starts to fall.

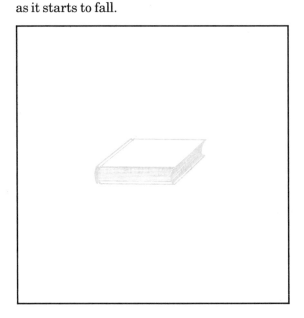

(i) Draw the force or forces acting on the book as it falls. Label the force or forces. (2 marks)
(ii) What is the acceleration of the book as it starts to fall? (1 mark)
(iii) The acceleration remains constant as the book falls to the floor. Sketch on the axes below a velocity – time graph for the falling book. (2 marks)

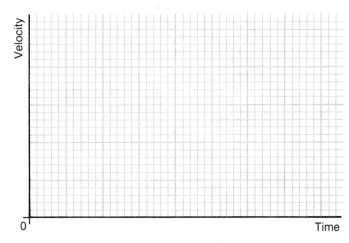

(Total: 11 marks)
(LEAG, Syll. A, Paper 2)

9 An instrument used by an astronaut has a mass of 25 kg on the earth. The gravitational field strength on the earth is 10 N/kg and on the moon it is 1.7 N/kg.
(a) What would be the mass of the instrument on the moon?
(b) What would be the weight of the instrument on the moon?

10 (a) In each of the following situations name the force which keeps the object moving in a circular path (i.e. say what provides the centripetal force).
(i) The moon moving round the earth.
(ii) A mass swung in a horizontal circle on the end of a string.
(iii) A car moving round a corner.
(b) In what direction does the centripetal force always act?

11 A girl of mass 45 kg jumps horizontally out of a small boat, with a velocity of 2 m/s. The mass of the boat is 15 kg. What is the change in the speed of the boat as the girl jumps out, and in what direction?

12 A 2 g bullet travelling at 100 m/s embeds itself in a stationary block of wood of mass 998 g which is at rest but free to move. What is the velocity of the block of wood as it starts to move?

13 A trolley is pulled back against an elastic band, as shown in the diagram.

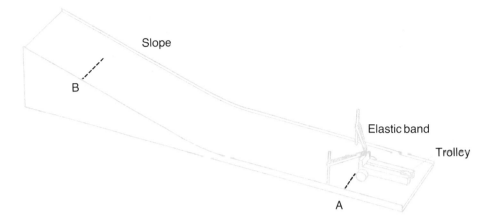

It is then released from position **A**. The trolley runs along the flat surface and then up the slope to **B**.
(a) The trolley comes to rest at **B**. What type of energy does it have at **B**? (1 mark)
(b) The trolley then runs back from **B**. It is stopped briefly by the elastic band stretching again. Using the axes shown, sketch a graph of *speed* against *time* for the trolley *from the moment it leaves* **B** until it comes to rest again. (3 marks)
(Total: 4 marks)
(MEG, Nuffield, Paper 2)

14 A ball of mass 0.2 kg is thrown vertically upwards from ground level and reaches a maximum height of 3.2 m. Ignoring air resistance calculate
(a) the potential energy the ball has gained when it is 3.2 m above the ground.
(b) the kinetic energy the ball gains in falling back to the ground.
(c) the speed of the ball just before it reaches the ground.
(d) the acceleration of the ball as it is falling.
(e) the force acting on the ball as it is falling.

15 The figure shows a fork-lift truck being used to stack crates.
(a) (i) Draw an arrow on the figure to show the direction of the force of gravity on the crate C.

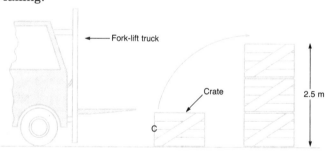

(ii) Each crate has a mass of 40 kg. The gravitational field strength is 10 N/kg. Use the equation

weight = mass × gravitational field strength

to calculate the weight of one crate.
(iii) Use the equation

work done = magnitude of force × distance moved in the direction of the force

to calculate the work done in putting crate C on top of the other crates.
(iv) Actually the fork-lift truck does more work than this in lifting the crate. Give **one** reason for the extra work. (8 marks)
(b) (i) What form of energy does the crate gain when it is lifted to the top of the stack?
(ii) How much energy is gained?
(iii) Add the two missing words in the following sentence:
'The Principle of Conservation of Energy states that energy cannot be _____ or _____; it simply changes from one form to another.' (4 marks)
(c) The crate slips and falls to the ground.
(i) What type of energy does it gain as it falls?
(ii) How much of this type of energy does it have just before it hits the ground?
(iii) What happens to most of the energy when the crate hits the ground? (3 marks)
(Total: 15 marks)
(NEA, Syll. B, Level P)

16 The table shown is based on information from the Highway Code. The figures apply to a saloon car in good driving conditions.

| Speed | Thinking distance | Braking distance |
|---|---|---|
| 10 m/s | 7 m | 8 m |
| 20 m/s | 14 m | 30 m |
| 30 m/s | 21 m | 68 m |

(a) The thinking distances depend on the driver's reactions.
 (i) How does the thinking distance depend on the car's speed? (2 marks)
 (ii) How much *thinking time* is the driver allowed according to the Highway Code? (2 marks)
(b) It takes the driver time to react (i.e. to register the need to brake) and apply the brakes. Once the brakes are applied it still takes the braking distance to stop the car.
 (i) According to the table what is the **total stopping distance** at 30 m/s? (1 mark)
 (ii) Two cars are following one another at 30 m/s. Suggest, with reasons, a **minimum** safe distance separating the cars. (2 marks)
 (iii) Estimate a maximum safe driving speed if fog has reduced visibility to 40 m. Explain your reasoning. (2 marks)
(c) A car of mass 600 kg is travelling at 20 m/s.
 (i) Calculate the kinetic energy of the car. (2 marks)
 (ii) What happens to this energy when the brakes are applied? (1 mark)
 (iii) What is the braking distance at this speed? Calculate the average braking force in bringing the car to rest from 20 m/s. (3 marks)
(d) If the speed of the car is doubled the braking distance increases approximately four times. By energy considerations or otherwise explain this change. (2 marks)
(Total: 17 marks)
(LEAG, Syll. B, Paper 3)

17 The electric motor, in the diagram, completely lifts the box in 5 s, at a steady speed. If the box weighs 100 N, calculate
 (i) the work done by the motor,
 (ii) the power of the motor.
(3 marks)
(WJEC, Paper 1)

18 There are four main forces on a rowing boat as it moves through water. These are *lift*, *weight*, *thrust* and *drag*. Their directions are shown on the diagram. *Thrust* is the force which drives the boat along. *Drag* is the force which opposes the motion of the boat.
 (a) Describe the motion of the boat when the thrust is greater than the drag. (1 mark)
 (b) What can you say about the sizes of *lift* and *weight*? (1 mark)
 (c) Some of the energy put into the rowing boat is wasted in the water. How? (1 mark)
The graph shows how the speed of the boat varies with the power input.

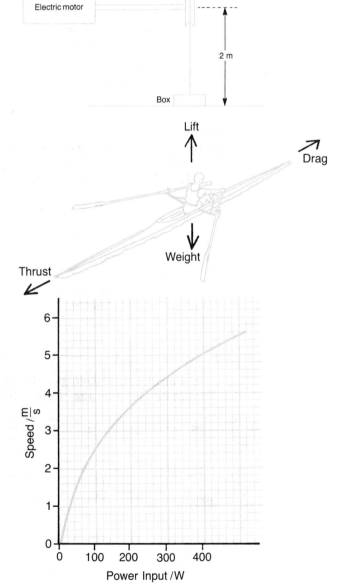

(d) From the graph find the power input when the speed of the boat is 4 m/s. (1 mark)
(e) The boat moves at 4 m/s for 6 minutes. Calculate the energy used.
(Use the equation *energy supplied = power × time*) (3 marks)
(Total: 7 marks)
(MEG, Syll. A, Paper 2)

19 An electric motor is used to lift a mass of 1 kg through a height of 3.2 m at a constant speed in a time of 8 s.
(a) Calculate
(i) the potential energy gained by the mass.
(ii) the useful work done by the motor.
(iii) the useful power output of the motor.
(b) When the mass reaches 2.45 m above the floor the string breaks.
(i) How much kinetic energy does the mass gain in falling to the floor?
(ii) What is the speed of the mass just before it reaches the floor?

20 (a) Explain the difference between a vector and a scalar quantity.
(b) Give two examples each of vector and scalar quantities.

21 Mary pulls her little brother Brian on a sledge. She pulls with a force 120 N and the friction force between the sledge and the ground is 20 N.
(a) What is the resultant accelerating force on the sledge? (1 mark)
(b) The graph shows the possible motion of the sledge as a result of the action of the accelerating force.
Using the information on the graph,
(i) find the speed of the sledge after 3 seconds, (1 mark)
(ii) find the speed of the sledge after 5 seconds, (1 mark)
(iii) The world record for the 100 m sprint is just under 10 seconds. Using this information, comment on your answer to (b)(ii).

(speed = $\frac{distance}{time}$) (3 marks)
(Total: 6 marks)
NEA, Syll. B, Level P)

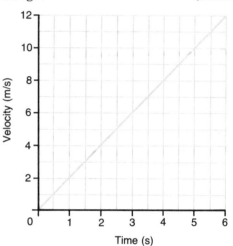

## Unit 4

22 The diagram below shows a uniform rod pivoted through its centre. A spring balance is attached to the rod 15 cm from the pivot. A force of 7.5 N is acting 40 cm from the pivot. The rod is balanced, and its weight may be neglected.

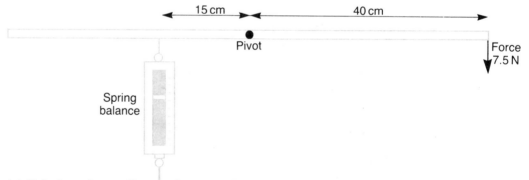

(a) Calculate the reading on the spring balance.
(b) State the value and direction of the force that the pivot exerts on the rod.

23

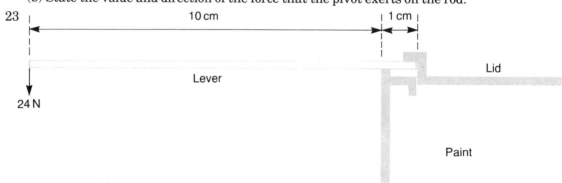

The diagram shows a lever being used to lift the lid from a tin of paint.
(a) On the diagram
(i) mark the position of the pivot P.
(ii) draw an arrow to show the direction of the force F, which the lever exerts on the lid.
(b) Calculate the moment of the force of 24 N about the pivot.
(c) Calculate the value of the force F which the lever exerts on the lid.
(d) What is the value of the force that the lever exerts on the pivot?
(e) If the force F is too small to lift the lid, suggest two changes which you might make to increase the value F.

24 It may be said that machines make work easier but do not make it any less. If you have, for example, a lifting job to do you may use a pulley system to do it, as in Figure 1.

(a) There are at least two good reasons for using a simple pulley to do this job. Let us see what they are.

(i) Calculate the work done in raising the load directly if it weights 300 N and has to be raised through a height of 20 m. (2 marks)

(ii) Calculate the work you would do in raising yourself through the same height if you weighed 600 N. (1 mark)

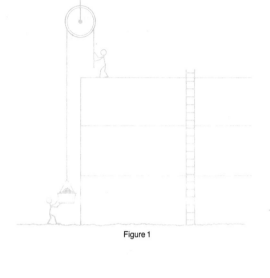

Figure 1

(iii) Can you now see one good reason for using the pulley? Explain your answer. (1 mark)

(iv) To raise the load yourself you have to climb the scaffolding ladders. What is the other good reason for using the pulley? (1 mark)

(b) If you used a pulley and a force of 350 N to raise the load at a steady speed, at what power would you work to raise it through 20 m in 35 s? (2 marks)

(Use the equation: Power = $\frac{\text{energy transferred}}{\text{time taken}}$)

(c) In reality, a block and tackle or compound pulley system would be used. The system shown in Figure 2 has four 'lifting' strings. Calculate
(i) the force in **one** lifting string if the total weight of the load and the lower pulleys is 400 N, (1 mark)
(ii) the fraction of the total weight represented by the 300 N load. (1 mark)
(Total: 9 marks)
(NEA, Syll. B, Level P)

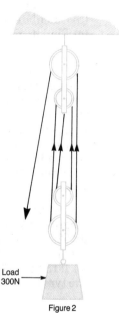

Load
300N

Figure 2

## Unit 5

25 The figure shows apparatus being used to find the **volume** of an iron bolt.

(a)   (b)   (c)

(a) (i) The apparatus is called a _____ cylinder.
   (ii) What liquid would you pour into the cylinder?
   (iii) What is the volume of the liquid in Figure (b)?
   (iv) In the third stage (Figure (c)) the iron bolt is lowered on a piece of string. Draw in the liquid level if the volume of the iron bolt is 20 cm³. (6 marks)
(b) (i) The mass of the bolt is 160 g. Name an instrument you would use to measure mass.
   (ii) Use the equation

$$\text{density} = \frac{\text{mass}}{\text{volume}}$$ to calculate the **density** of iron. (4 marks)
(Total: 10 marks)
(SEG, Syll. A, Level P)

26 A graduated cylinder contains 60 cm³ of oil. When a lump of ice was added the level rose to 90 cm³ and when the ice melted the level fell to 87 cm³.

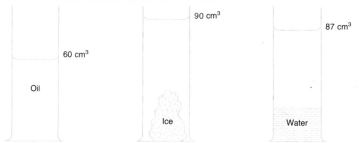

(a) Determine from the diagrams
(i) the volume of ice   (ii) the volume of water   (2 marks)
(b) If the density of water is 1 g/cm³, calculate
(i) the mass of water   (ii) the density of ice   (3 marks)
(c) Suggest a suitable value for the density of the oil. (1 mark)
(Total: 6 marks)
(WJEC, Paper 2)

27 (a) Calculate the surface area of a cube of side 2 cm.
(b) Calculate the volume of the same cube.
(c) Find its mass if it has a density of 8 g/cm³.

## Unit 6

28 The figure shows side and front views of a car tyre in contact with the road.
One tyre company stated that the area of the tyre in contact with the road was about the same as the area of the sole of one of your shoes.

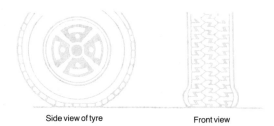

Side view of tyre    Front view

(a) Describe how you would estimate the area of the sole of your shoe. (You may use a diagram if you wish.) (4 marks)
(b) The car weights 12 000 N. What is the force acting on one tyre if the weight of the car is evenly distributed amongst the tyres? (3 marks)
(c) If the area of contact of the tyre is 80 cm² (0.008 m²), calculate the pressure of the air in the tyres. (4 marks)

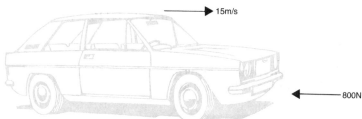

(d) The car is travelling at a steady speed of 15 m/s.
(i) Calculate the distance moved in 10 s. (4 marks)
(ii) The total resisting force (friction and air resistance) is 800 N. Calculate the work done by the car in 10 s in overcoming this resisting force. (4 marks)
(iii) Calculate the power developed at the driving wheels. (4 marks)
(Total: 23 marks)
(NEA, Syll. A, Level P)

29 A hovercraft has a mass of 10 000 kg. It hovers at a constant height above the ground.
(a) Calculate the weight of the hovercraft.
(b) What is the value of the upward force exerted by the air cushion?
(c) The hovercraft has a rectangular shape of length 20 m and width 5 m. Calculate the pressure excess (above atmosphere) of the air in the cushion under the craft.
(d) The hovercraft accelerates horizontally at 2.5 m/s². Calculate the horizontal force exerted by the driving propeller. You may ignore air resistance.

30 On what factors does the pressure under the surface of a liquid depend?

## Unit 7

31 The diagram shows a hydraulic press used to crush waste paper. The area of piston Q is ten times that of piston P. Pressure is applied to the oil by pressing up on piston P.
(a) How does the pressure on the oil above piston P compare with that above piston Q? (1 mark)
(b) (i) How does the downward force on piston Q compare with the upward force on piston P? (1 mark)

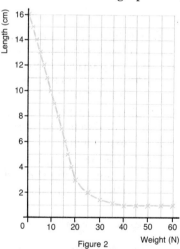

(ii) Explain your answer. (1 mark)
(c) Why is oil in this machine more suitable than air? (1 mark)
(d) When piston P moves up 20 cm, explain why piston Q moves down 2 cm. (3 marks)
(Total: 7 marks)
(MEG, Syll. A, Paper 2)

32 Jim takes a spring off an old mattress and decides to experiment with it at school. He sets up the apparatus shown in Figure 1 and loads the spring with 5 N weights, recording the length of the spring at each loading. He obtains the results shown in the graph in Figure 2.

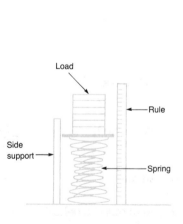

Figure 1                                Figure 2

(a) Over which range of load values can the behaviour of the spring be described by Hooke's law? Give a reason for your answer. (2 marks)
(b) Why would Jim test the spring for compression rather than for extension? (1 mark)
(c) What has happened to the spring when it has a length of 1 cm? (1 mark)
(d) If Jim's mother put a packing case on the mattress and it weighed 300 N, what would be the **average** compression of the springs if the case lay directly over 30 of them? (3 marks)
(Total: 7 marks)
(NEA, Syll. B, Level P)

33 In an experiment to investigate how the length of a strip of rubber changed when the strip was loaded the following table of readings was produced.

| Length/cm | 10 | 12 | 17 | 20 | 25 |
|---|---|---|---|---|---|
| Load/N | 0 | 1 | 3.5 | 5 | 7.5 |

(a) On graph paper, plot a graph of the length of the rubber strip against load. (2 marks)

(b) Use your graph to find
(i) the length of the strip for a 2.7 N load, (1 mark)
(ii) the extension produced by a 6 N load, (1 mark)
(iii) the load that produces a 2.5 cm extension. (1 mark)
(Total: 5 marks)
(WJEC, Paper 2, Section B)

## Unit 8

34 The model shown in the diagram is sometimes used to represent the arrangement of molecules in a solid.
(a) What do the springs represent?
(b) How may the model be used to show what happens when a solid is heated?
(c) State how the movement of molecules in a liquid differs from the movement of molecules in a solid.
(d) Why is energy needed to melt a solid into a liquid at the same temperature?

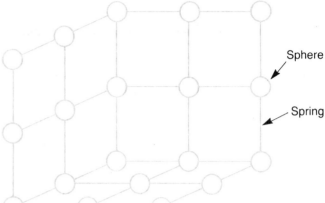

35 This question is about the ways in which molecules move. You are given three descriptions:
One description refers to the molecules of a solid.
One description refers to the molecules of a liquid.
One description refers to the molecules of a gas.
Write either **solid** or **liquid** or **gas** at the end of each description.
(i) close packed, moving to and fro about fixed positions _____
(ii) widely separated, moving freely and randomly _____
(iii) close packed, jostling and changing places with neighbouring molecules _____

(Total: 3 marks)
(LEAG, Syll. A, Paper 2)

36 The figure shows apparatus used to observe the behaviour of smoke particles in air.
(a) Why is light shone into the container? (1 mark)
(b) Why are smoke particles very suitable for use in this experiment? (2 marks)
(c) Describe what you would see when looking through the microscope into the smoke cell. (3 marks)

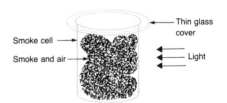

(d) What does the experiment tell you about the behaviour of the air molecules in the cell? (2 marks)
(e) What difference, if any, would be seen in the motion of the smoke particles if a weaker light was used? (1 mark)
(Total: 9 marks)
(NEA, Syll. B, Level P)

37 The ampoule is placed inside the rubber tubing and then broken using pliers. The tap is then opened to let bromine run into the tall tube. The bromine vaporizes and the colour spreads up the tube.
(a) Name the process that is demonstrated by this experiment. (1 mark)
(b) Describe what you would see
(i) about 10 s after opening the tap;
(ii) about 10 minutes after opening the tap. (2 marks)
(c) (i) The average speed of a bromine molecule is 200 m/s. Calculate the **total** distance travelled by a typical bromine molecule in 10 s. (2 marks)
(ii) Draw a diagram to illustrate the path of a typical bromine molecule. With the help of the diagram explain why no bromine will be observed at the top of the tube after 10 s. (4 marks)

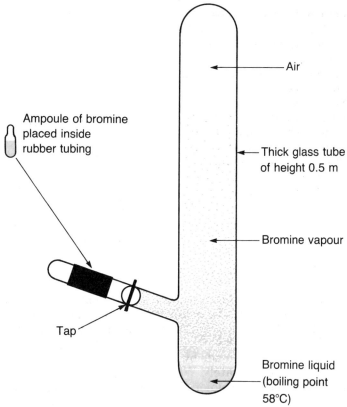

(iii) How could this apparatus be used so that the bromine reaches the top of the tube in a fraction of a second? Explain. (2 marks)

(d) Evaporation requires energy. The tube is **not** heated and yet a lot of bromine evaporates. The remaining liquid bromine becomes colder.
Explain the process of evaporation in terms of the behaviour of bromine molecules.
(4 marks)
(Total: 15 marks)
(LEAG, Syll. B, Paper 3)

38 Some perfume is spilled in a corner of a room in which the doors and windows are closed to stop draughts. The scent slowly spreads by *diffusion* to all parts of the room.
(a) By writing about molecules, explain how *diffusion* occurs. (2 marks)
(b) Why does the scent spread slowly even though the molecules move fast? (3 marks)
(Total: 5 marks)
(MEG, Syll. A, Paper 2)

## Unit 9

39 Figure 1 is a diagram of a bar and gauge which is used to show expansion.
(a) (i) What must be done to the bar in order to make it expand?
(ii) What piece of apparatus is used to make the bar expand?
(iii) Name a suitable material for the handle and give a reason for your choice.
(4 marks)
(b) A bimetallic strip made of iron and copper is shown in Figure 2(a). The strip is at room temperature. Copper expands more than iron for the same temperature rise. Complete Figure 2(b) to show what happens to the bimetallic strip when it is cooled. (2 marks)

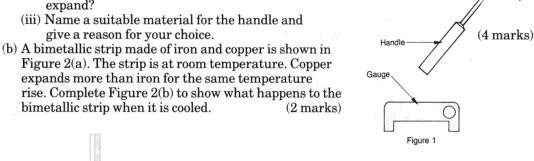

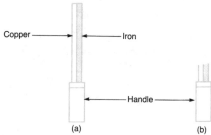

Figure 2

(c) A bimetallic strip can be used in a fire alarm when connected in a circuit like the one in Figure 3.
 (i) Explain how the alarm works when a fire starts.
 (ii) Write down another use for a bimetallic strip.  (4 marks)
(d) Figure 4 shows tapered joints between two sections of railway line. Explain why the joint is like this.  (3 marks)
(Total: 13 marks)
(NEA, Syll. A, Level P)

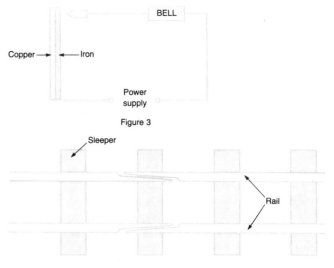

40 (a) Explain in terms of the kinetic theory why a metal bar expands on heating.
(b) Two metal plates may be riveted together by placing a white hot rivet in the hole through the two plates. The head of the rivet is then hammered flat. Explain why the plates are held more tightly together when the rivet was cooled.
(c) A bimetallic strip is made by riveting together a strip of iron and a strip of brass so that they cannot move separately. When heated, brass expands more than iron. Draw a diagram to show the shape of the bimetallic strip after it has been heated. Label the metals.

41 You are trying to measure the boiling point of water using a thermometer as shown in the figure.
 (a) (i) What does a thermometer measure?
     (ii) What liquid is found in this thermometer?
     (iii) What liquid would be used in a thermometer to measure a boiling point of about −50°C?  (3 marks)
(b) What is the boiling point of water
 (i) in degrees Celsius,    (ii) in kelvin?  (2 marks)
(c) A nurse may use a clinical thermometer to measure a patient's temperature.
 (i) Give **two** reasons why this thermometer is more suitable than the one used in part (a).
 (ii) For a healthy person the instrument will read 37°C. Change 37°C to kelvin.  (4 marks)
(d) Write down the name of an instrument that could be used to measure up to 1500°C as in a furnace.  (1 mark)
(Total: 10 marks)
(NEA, Syll. A, Level P)

## Unit 10

42 More air is pumped into a bicycle tyre. Explain why the pressure becomes bigger.
(2 marks)
(MEG, Nuffield, Paper 2)

43 The figure illustrates an apparatus in which a fixed mass of air was compressed in a calibrated syringe, which was approximately half full of air at atmospheric pressure and a temperature of 17°C.

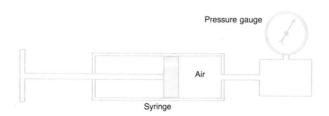

Corresponding values of volume and pressure of the trapped air are shown in the table.

| pressure (kPa) | 50 | 60 | 75 | 90 | 105 | 120 |
|---|---|---|---|---|---|---|
| volume (m³) | 0.00048 | 0.00040 | 0.00032 | 0.00027 | 0.00023 | 0.00020 |
| $\frac{1}{volume}$ (m⁻³) | | 2500 | | 3704 | | 5000 |

(a) Complete the table by calculating values for $\frac{1}{volume}$. Some of the values have been entered for you. (3 marks)

(b) Plot a graph of pressure on the y-axis against $\frac{1}{volume}$ on the x-axis. (8 marks)

(c) What relationship between pressure and volume of the trapped air can be deduced from your graph? Explain your answer. (3 marks)

(d) If the temperature of the air was increased to 27°C, what volume would be occupied by the air at a pressure of 100 kPa? (6 marks)

(Total: 20 marks)
(NEA, Syll. A, Level Q)

44 Some air at a pressure of 1 atmosphere and a temperature of 27°C occupies a volume of 1 litre.
(a) The air is heated to 177°C at constant pressure. What will be its new volume?
(b) The air is now cooled to its original temperature keeping the volume constant. Calculate its final pressure.

45 (a) State what you understand by the statement absolute zero is approximately −273°C.
(b) Why is absolute zero thought to be a better zero than 0°C?

46 The metal can has a tight fitting lid. The can is heated.
  (i) What happens to the motion of the air molecules in the can when the temperature is increased? (2 marks)
  (ii) What happens to the pressure in the can when the temperature increases? (2 marks)
  Explain your answer in terms of the behaviour of air molecules. (3 marks)
  (iii) The lid eventually flies off. Explain this by comparing the behaviour of the air molecules inside and outside the can. (2 marks)

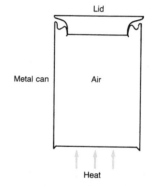

(b) A teacher demonstrated this experiment but to make it work 'better' the teacher put a little water in the bottom of the can before heating.
Explain how this increases the pressure in the can. Refer to the water molecules in your answer. (2 marks)

(c) Aerosol cans often come with a danger warning of the type:
**'Caution; pressurized container; protect from sunlight and do not expose to temperatures exceeding 50°C.'**
Suggest reasons for the caution. (2 marks)

(Total: 13 marks)
(LEAG, Syll. B, Paper 2)

## Unit 11

47 When 1 kg of water was heated for 5 minutes with an immersion heater the water's temperature rose by 30°C.
To the questions that follow answer **either**: greater than 30°C; **or** less than 30°C; **or** equal to 30°C.
What would be the temperature rise if the same heater heated
(i) 2 kg of water for 9 minutes?
(ii) 1 kg of paraffin for 5 minutes (given that the heat capacity of the paraffin is less than that of the water)? (2 marks)
(WJEC, Paper 1)

48 This question is concerned with a proposed system in which solar energy collected in roof-top panels is stored as thermal energy in a large tank of water beneath a house. The energy is extracted by a heat exchanger and used for space heating in the house. The figure (overleaf) shows a typical arrangement.

**152** *Short answer questions*

(a) The average power from the sun reaching the solar panel during June is 300 W/m². If the area of the panel is 75 m², how much energy per second will be falling on the panel during June? (2 marks)

(b) On the axes drawn below, sketch a graph to show how you think the energy arriving at the solar panel per second will vary according to the time of year. You can assume that the house is in the northern hemisphere. Label this graph A. (2 marks)

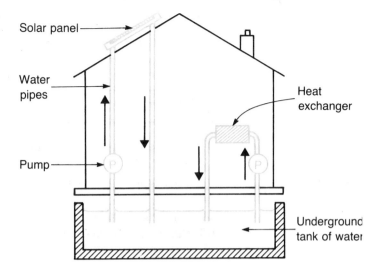

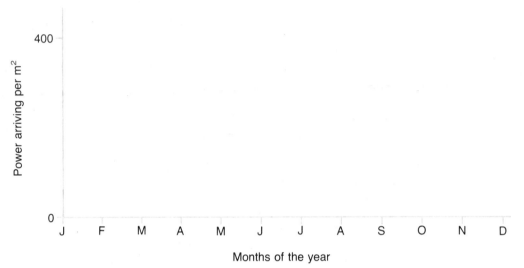

(c) On the same axes, draw a curve to show how the energy demand may vary with the month of the year. Label this graph B. (2 marks)

(d) From your graph, when will the temperature of the water in the underground tank be at its maximum value? Explain your answer. (3 marks)

(e) If the volume of water in the tank is 50 m³ and it takes 4200 J of energy to increase the temperature of 1 kg of water by 1 K, how much extra internal energy would be stored in the water if its temperature rose by 25 K? (Density of water = 1000 kg/m³) (3 marks)

(f) If the average power falling on the panel is 300 W/m², approximately how many days would it take to increase the temperature of the water by 25 K? The area of the panel is 75 m². (4 marks)

(g) Give **three** reasons why your answer to part (f) will be too low. (3 marks)

(Total: 19 marks)
(NEA, Syll. B, Level Q)

## Unit 12

49 (a) Water gives out heat as it freezes at 0°C. Where does the heat come from?
(b) What quantity of heat must be removed from 500 g of water at 0°C to change it to ice at 0°C? (Specific latent heat of fusion of ice = 334 J/g)

50 (a) Why is a scald from steam much worse than one from boiling water?
(b) Why can a bottle of milk be kept cool in hot weather by covering it with a wet cloth?

51 (a) Heat is required to convert water to steam at 100°C. What happens to this heat?
(b) A 3 kW electric kettle is left on for 3 minutes after the water starts to boil. What mass of water is boiled off in this time? (The specific latent heat of vaporization is $2.26 \times 10^6$ J/kg.)

52 (a) Use the molecular theory to explain why evaporation causes cooling. (2 marks)
(b) State **two** ways by which the rate of evaporation of a liquid may be increased. (2 marks)
(c) State **two** ways in which boiling is different from evaporation. (2 marks)

(Total: 6 marks)
(WJEC, Paper 2)

## Unit 13

53 Complete the following sentences:
   (a) Heat travels by _____ in solids.
   (b) Heat travels mainly by _____ in liquids and gases.
   (c) Heat travels mainly by _____ in space.

54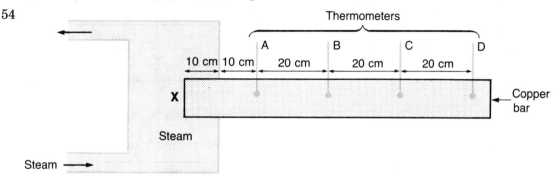

A solid copper bar has four holes into which are placed four identical thermometers A, B, C and D. The end X of the bar is heated by steam at 100°C.
After 15 minutes the thermometers all give steady readings:
A = 80°C;   B = 55°C;   C = 40°C;   D = 30°C
   (a) (i) Plot a graph of temperature against distance along the bar measured from X.
       (ii) Use your graph to find the temperature at a point 30 cm from end X. (4 marks)
   (b) (i) Name the way in which heat passes along the bar.
       (ii) Name **two** ways in which the bar loses heat to the surroundings.
       (iii) Describe how **one** of these may be reduced. (3 marks)
                                                           (Total: 7 marks)
                                                           (WJEC, Paper 2)

55 A beaker of water is heated, as in the diagram, and the movement of the water is shown by using a coloured dye.
   (a) On the diagram show the direction of movement of the hot water
   (b) Explain why the hot water moves in the way you have shown. (3 marks)
       (WJEC, Paper 1)

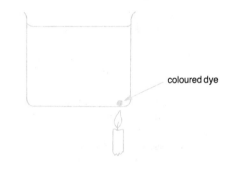

56 The diagram shows some of the main energy losses from a house.

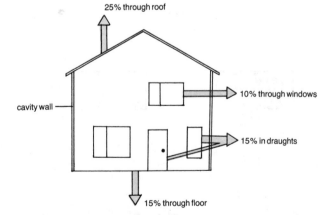

   (a) Energy is also lost through the cavity walls.
   (i) What percentage is this of the total energy loss? (1 mark)
   (ii) How can the energy loss through the cavity walls be made less? (1 mark)
   (b) The owner decides to improve the insulation of the house by fitting double glazing **or** insulating the roof. Which of these two methods saves more in heating costs? Explain your answer. (2 marks)
   (c) Explain how each of the following reduces energy loss:
   (i) fitting aluminium foil behind radiators (1 mark)
   (ii) blocking off unused fire places and their chimneys (1 mark)
                                                           (Total: 6 marks)
                                                           (MEG, Syll. A, Paper 2)

57 This question is about the importance of saving energy in the home and how it can help the country as a whole.
(a) A customer of British Gas had double-glazing installed in January 1987. Table 1 shows the readings of her gas meter for the dates shown.

| Date | 1986 | | 1987 | |
| --- | --- | --- | --- | --- |
| | 1 January | 31 December | 1 January | 31 December |
| Reading (units) | 6517 | 7665 | 7665 | 8616 |

Table 1

If each unit costs 38p, how much money did she appear to have saved in 1987 compared with 1986? (4 marks)
(b) Table 2 below shows some information about home insulation.

| Type of insulation | Approximate cost (£) | Pay-back time (yr) | % of thermal energy saved |
| --- | --- | --- | --- |
| loft | 180 | 5 | 10 |
| double glazing | 5000 | 70 | 25 |
| cavity wall | 700 | 10 | 20 |

Table 2

From Table 2,
(i) choose which insulation method would cost the least per pay-back year, (pay-back time is the time it takes for the savings to amount to the original cost of installing the insulation), (2 marks)
(ii) state which method would result in the greatest percentage saving of energy. (1 mark)
(c) Besides those listed, there are other smaller-scale methods of insulation against heat loss in the home. Name **two**. (2 marks)
(d) (i) If the average cost of insulating each uninsulated home in Britain is £1000, and there are 5 million such dwellings, how much would it cost a Government to insulate them all? (1 mark)
(ii) If the saving per year is £500 million, what would be the pay-back time, in years, for the country? (1 mark)
(e) There are different sorts of double-glazing, with different air gaps and different thicknesses of glass. Suggest **one** reason in each case for limiting the size of the air gap and the thickness of the glass used. (2 marks)
(Total: 13 marks)
(NEA, Syll. B, Level P)

58 (a) The diagram shows the cross-section of the three types of window.

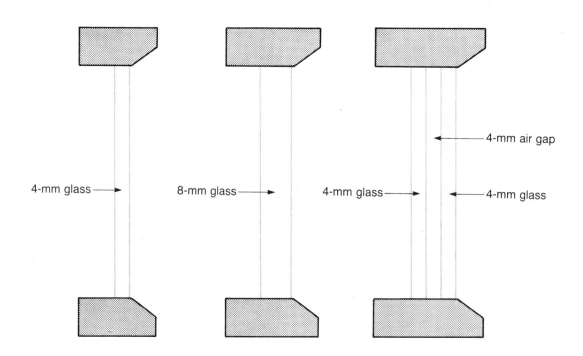

Energy is lost through a 4-mm glass window at the rate of 120 W.
  (i) If the 4-mm glass is replaced by 8-mm glass, will the energy loss be the same, greater or less? Give a reason to support your answer. (2 marks)
  (ii) The energy loss through the double-glazed window is much less than through either of the other two. Why is this? (2 marks)
  (iii) What other advantage does the double-glazed window have? (1 mark)
(b) The next diagram shows three views of a **cavity wall**. This has a larger air gap than the double-glazed window and convection currents are able to circulate in the cavity. The right-hand diagram shows a cavity wall which has been filled with insulating foam.

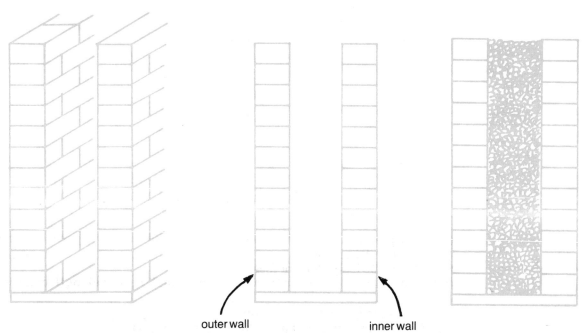

  (i) Draw possible convection currents on the middle diagram. (2 marks)
  (ii) Cavity wall insulation traps pockets of air. What effect does this trapped air have on heat transfer by convection? Give a reason to support your answer. (2 marks)

(Total: 9 marks)
(LEAG, Syll. A, Paper 2)

## Unit 14

59 (a) Waves may be either transverse or longitudinal. Why are longitudinal waves so named?
(b) Which of the following are transverse waves and which are longitudinal?
Waves on water, sound waves, radio waves, light waves.

60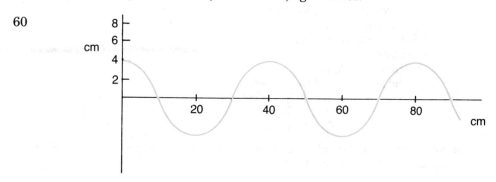

The diagram above shows a wave travelling along a length of rope.
(a) Using the scales on the given axes determine
(i) the amplitude of the wave,    (ii) the wavelength of the wave.
(b) If the frequency of the wave is 8 Hz, calculate the speed of the wave in **m/s**. (3 marks)
(WJEC, Paper 1)

61 (a) A ripple tank is used to observe the ripples made by a point source. One ripple is produced every 0.1 second. The first few ripples produced are shown in the diagram overleaf. The ripples hit a metal barrier a short time later.

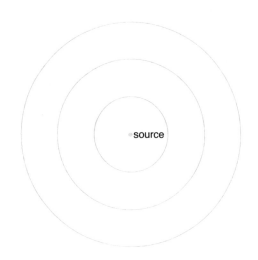

(i) The ripples are circles. What does this tell you about the way in which the waves travel in different directions? (1 mark)
(ii) What is the **frequency** of these waves? (2 marks)
(iii) By taking measurements from the diagram find the **wavelength** of these waves. (1 mark)
(iv) Calculate the speed of the waves. (2 marks)
(v) What happens to the ripples when they hit the barrier? (1 mark)
(vi) Add **three** more ripples with the same wavelength to the upper part of the diagram. (Freehand curves will do if you do not have a pair of compasses.) (3 marks)
(Total: 10 marks)
(LEAG, Syll. B, Paper 2)

62 A vibrating source in a ripple tank produces a series of equally spaced circular ripples which strike a straight reflector.
(a) Complete the diagram to show how crests 4 and 5 are reflected. (2 marks)
(b) If the velocity of the waves is 96 mm/s and the source is 36 mm from the reflector, calculate
(i) the wavelength
(ii) the frequency of the waves. (3 marks)
(Total: 5 marks)
(WJEC, Paper 2)

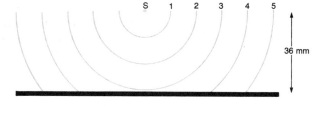

63 Water waves approach a harbour mouth as shown in the diagram.

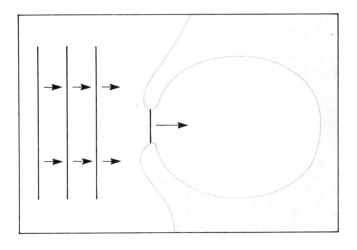

(a) Complete the diagram to show the wave pattern in the harbour. Ignore reflections from the harbour walls.
(b) What is the frequency of this effect?
(c) The frequency of the incoming waves increases. What effect does this have on the wavelength of the waves?
(d) How would the change in (c) affect your answer to (a)?

## Unit 15

64

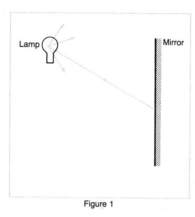
Figure 1

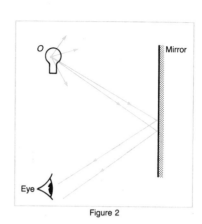
Figure 2

(a) Figure 1 represents a ray of light from a lamp shining onto a mirror. Show how the ray leaves the mirror after reflection. (2 marks)
(b) Figure 2 represents two rays of light leaving the mirror and entering an eye. These rays come from $O$, the dot at the centre of the lamp.
  (i) Use a letter $I$ to mark the image position on the diagram. The image position is where the rays reaching the eye appear to come from. (3 marks)
  (ii) This image is a virtual image. What is meant by the term *virtual*? (2 marks)
  (iii) Mark, on Figure 2, a normal to the mirror for one of the two reflected rays and label the angle of reflection for this ray. (2 marks)
(c) Figure 3 shows the mirror in a new position. The rays of light still come from $O$. Show with a letter $I$ where the light reaching the eye now appears to come from. Show on the diagram how you decided on the position of $I$. (3 marks)
(Total: 12 marks)
(LEAG, Syll. A, Paper 2)

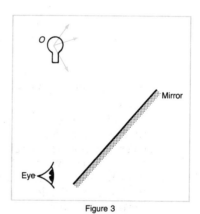
Figure 3

65 The diagram shows a periscope which may be used to see behind when towing a caravan.
Only one mirror has been drawn in.
(a) Draw the second mirror in the correct position.
(b) Draw a ray of light from the top of the object to the eye.
(c) Draw a ray of light from the bottom of the object to the eye.
(d) What is the main disadvantage of this type of periscope?

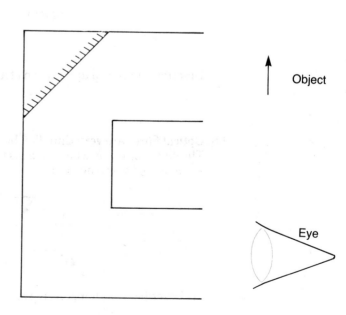

## Unit 16

66 The diagram shows three rays of light from one point on the bottom of a swimming bath reaching the surface.

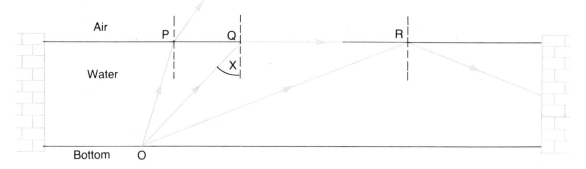

(a) What name is given to
(i) the bending of the ray at P, (ii) the angle of incidence marked X?
(b) Explain why the ray bends in the direction shown at
(i) the point P, (ii) the point R.

67 (a) Figure 1 shows a ray of light passing through a transparent material. Name the angles marked X and Y. (2 marks)

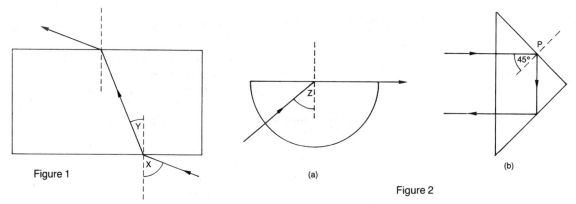

(b) Figure 2(a) and 2(b) each show a ray of light entering and leaving a transparent plastic block.
(i) In Figure 2(a) what name is given to the angle Z? (1 mark)
(ii) For Figure 2(b),
(A) name the type of reflection at the point P,
(B) explain why this reflection took place,
(C) where is this effect used on a bicycle? (5 marks)
(c) The type of reflection used in Figure 2(b) is used to pass a beam of light along a solid glass fibre. This is shown in Figure 3.

Figure 3

Describe a practical application of a single flexible fibre or a bundle of flexible fibres.
(4 marks)
(Total: 12 marks)
(NEA, Syll. A, Level P)

68 Optical fibres are very thin, flexible rods used for carrying light.
The diagram shows what happens to a ray of light AN which enters the end of an optical fibre at an angle *i* to the normal.

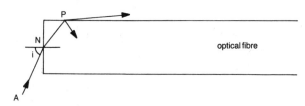

(a) Describe what happens to this ray at N. (2 marks)

(b) Describe what happens to this ray at P. (2 marks)

The diagram below shows a different ray BN entering the end of the optical fibre at a much **smaller** angle $i$ to the normal.

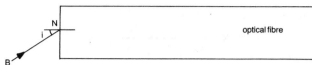

(c) On the diagram draw the path of this ray into and through the fibre. (2 marks)
(d) State **one** use of optical fibres. (1 mark)
(Total: 7 marks)
(MEG, Syll. A, Paper 2)

69 Three of the boxes below contain lenses. The shape of the items in the boxes is not shown. You have to work out from the ray paths what each box contains.

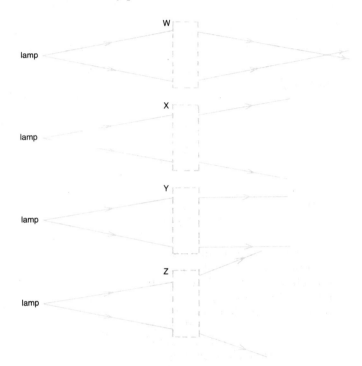

(a) Which **two** boxes contain *converging* lenses? (2 marks)
(b) Which **one** box contains a *diverging* lens? (1 mark)
(c) In the boxes sketch the shapes of the lenses. (3 marks)
(d) What could the other box contain? Give a reason for your answer. (2 marks)
(e) The items in the boxes are made from glass or perspex. What property of glass or perspex is responsible for the changes in direction of the rays as they pass through the boxes?
(2 marks)
(Total: 10 marks)
(LEAG, Syll. B, Paper 2)

70

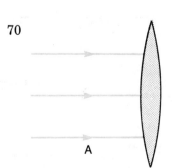
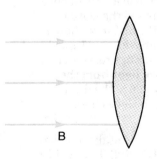
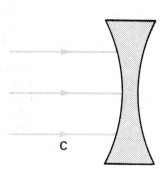

(a) In each of the parts of the figure, drawn to the same scale, rays of light are shown striking a lens. Copy the drawings and show in each case possible directions of the rays when they come out of the lens.
(b) Lens D in the figure overleaf forms an image I of a very small bright object O.
(i) If the small object is raised 2 cm above the line XY, where is the new image position?

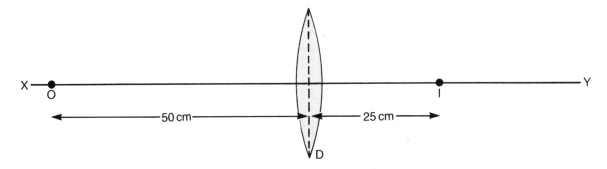

(ii) Draw a sketch to show how the lens forms the image of the object in its new position. You should show the path of two rays from the object.
(iii) If the object is moved a short distance further away from the lens, what happens to the image?
(c) Draw a suitable ray diagram to show how such a lens can be used as a magnifying glass. The rays in your diagram should show the direction in which the light waves travel.

71 (a) In a thunder storm, lightning is seen before the thunder is heard. What does this tell you about light? (2 marks)
(b) Explain why **blue** light is refracted more than red. (2 marks)

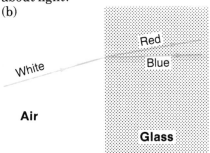

(c) A camera is focused to take a clear picture of a small lamp. Complete the paths of the **three** rays, to show how the lens forms the image of the lamp on the film.
(2 marks)
(Total: 6 marks)
(MEG, Nuffield, Paper 2)

72 A man finds that at a distance of 25 cm the words in a book look blurred.
(a) From what eye defect does the man suffer? (1 mark)
(b) In what direction should he move the book to be able to see the words clearly? (1 mark)
(c) What spectacle lens is needed to allow him to see the words clearly from the distance shown in the figure above? (1 mark)
(d) On the lower figure, sketch how the rays of light from the book would produce the final image, using the lens chosen in (c). (3 marks)

(Total: 6 marks)
(NEA, Syll. B, Level P)

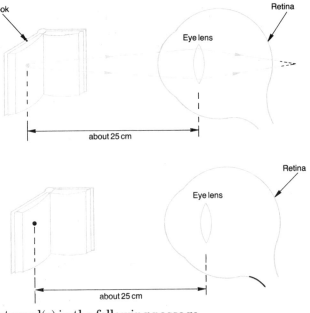

73 Fill in the blank spaces with the correct word(s) in the following passage.
A person suffers from long sight because his eye lens is too _____ . When looking at a close object the image is brought to focus _____ the retina. This fault may be remedied by using spectacles with a _____ lens so that the image is now formed _____ the retina.

74 Fill in the blank spaces with the correct word(s) in the following passage.
   A person suffers from short sight because his eyeball is too _____. When looking at a distant object the image is brought to focus _____ the retina. This fault may be remedied by using spectacles with a _____ lens so that the image is now formed _____ the retina.

## Unit 17

75 The diagram illustrates the diffraction of light waves passing through a narrow slit.
   (a) How could the spreading of the waves be increased?
   (b) Does the wavelength of the light increase, decrease or remain the same as the light passes through the slit?
   (c) What change, if any, would you notice in the diffraction caused by the slit, if the wavelength of the incoming light were greater?

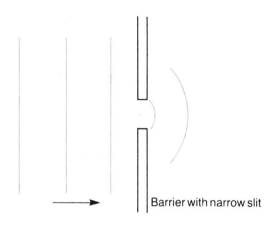

76 To observe Young's fringes a blackened microscope slide with two slits close together is placed between a lamp and a screen, in a dark room, as shown in the diagram. Bright and dark bands (fringes) are seen on the screen.

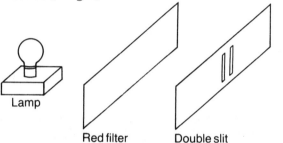 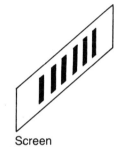

   (a) What property of light does this experiment demonstrate?
   (b) (i) If light from the two slits arrives at the screen 'in step', do bright or dark bands appear?
       (ii) If the light reaches the screen 'out of step', what sort of bands appear on the screen?
   The pattern on the screen can be explained by considering the difference in path length as shown below.

   (c) What type of band will be seen on the screen in each of the following cases?
   (i) The paths are equal in length.
   (ii) The paths are different by one wavelength.
   (iii) The paths are different by half a wavelength.

   (d) Blue light has a shorter wavelength than red light. What would happen to the fringe separation if a blue filter were used instead of a red one?
   (e) What would happen to the fringe separation if the slits were replaced by two slits which were closer together?

77 The table gives some examples of electromagnetic waves (not to scale) and their wavelengths. They all have the same speed of 300 000 000 m/s.

| Wave | Radio | Microwaves | Infrared | Visible light | Ultraviolet |
|---|---|---|---|---|---|
| Wavelength | 100 m | 0.1 m | 1 µm | 3/10–7/10 µm | 1/100 µm |

   (a) Which of the examples given has the longest wavelength and which the shortest?
   (2 marks)
   (b) What is the frequency of radio waves? (frequency = speed/wavelength) (2 marks)
   (c) Microwaves are examples of electromagnetic waves and can be used for cooking if they are of the correct frequency (2 500 000 000 hertz). They reflect well off metals and can pass through most non-metals.
   (i) Why do you think that the walls of a microwave oven are made of steel? (2 marks)

(ii) Microwave ovens are assumed to cook more efficiently than gas or electric ovens. Give a reason why this might be thought to be a reasonable assumption. (2 marks)
(iii) Water molecules inside food absorb the energy of the microwaves. What is the microwave energy changed to inside the food? (1 mark)
(iv) The speed of the microwaves is reduced when inside the food.
   (A) What happens to the wavelength of the waves as a result of this change? (1 mark)
   (B) What happens to their frequency? (1 mark)
(Total: 11 marks)
(NEA, Syll. B, Level P)

78 (a) Study the following list of types of wave.

  **A** sound waves     **C** X-rays     **E** water waves in a pond
  **B** light waves     **D** radio waves

Choose letters from the list to answer the following questions.
(i) Which can travel in a vacuum? (Give **three** answers.)
(ii) Which are electromagnetic? (Give **three** answers.)
(iii) Which is longitudinal? (Give **one** answer.)
(iv) Which can have wavelengths of hundreds of metres? (Give **one** answer.) (6 marks)
(b) Sunil draws three diagrams in his notebook to show waves being **reflected**, **refracted** and **diffracted**. The diagrams are shown below. Name of the wave property he is trying to draw in each diagram. (3 marks)

(Total: 9 marks)
(LEAG, Syll. A, Paper 2)

79 Light and infrared radiation are both examples of electromagnetic waves.
(a) State **two** differences between light and infrared radiation.
(b) Describe briefly **one** practical use of infrared radiation. (3 marks)
(WJEC, Paper 1)

## Unit 18

80 A ship fires a gun and the crew hear an echo from the cliff face opposite 5 s later. If the velocity of sound is 300 m/s, what is the distance of the ship from the cliff?

81 A fishing boat is using a sonar device which sends short pulses of sound vertically downwards. It receives an echo back after 0.5 seconds.
(a) How is the echo produced? (1 mark)
(b) Calculate how far the sound pulse travels in 0.5 seconds. Use the fact that the speed of sound in sea water is 1500 m/s. Show your working. (3 marks)
(c) How deep is the water under the boat? (1 mark)
(Total: 5 marks)
(MEG, Syll. A, Paper 2)

82 (a) Two students wish to study the wave forms produced by various sounds using a cathode ray oscilloscope (C.R.O.). Draw a sketch of the experimental arrangement they would use. (2 marks)
(b) Using a tuning fork they obtained the following trace.

Without adjusting the oscilloscope but using different sources of sound the following traces were obtained.
(i)                                           (ii)

How do the sounds heard from the new sources compare with the original sound from the tuning fork? (2 marks)
(Total: 4 marks)
(WJEC, Paper 1)

## Unit 19

83 The diagram shows two bar magnets suspended by similar threads.
(a) One end of magnet Q is brought near to the S pole of magnet P. The magnets attract. Label the S pole of magnet Q.
(b) If the other end of magnet Q is brought near to the S pole of magnet P, what will happen?
(c) State what you understand by the expression 'magnetic field'.

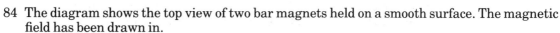

84 The diagram shows the top view of two bar magnets held on a smooth surface. The magnetic field has been drawn in.
(a) Describe the field at point X.
(b) Y and Z are plotting compasses. Draw an arrow in each circle to show the direction in which the compass needle points.
(c) What will happen to the magnets if they are released?
(d) One of the magnets is turned round as shown below. Draw the magnetic field lines within the dotted box.

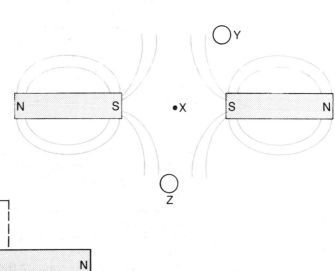

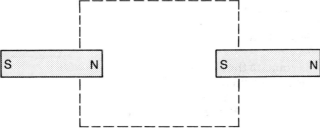

## Unit 20

85 (a) State two ways in which an electroscope may be charged by using a polythene rod.
(b) Does the cap of the electroscope receive the same or different signs of charge in the two cases?

86 A gold leaf electroscope is charged positively. An insulated conductor is brought up close to but not touching the cap of the electroscope. What change will be observed in the deflection of the leaf of the electroscope if the conductor is (a) positively charged, (b) negatively charged, (c) uncharged?

## Unit 21

87 Two resistors are in series with an ammeter and a 5 V battery.

The ammeter reads 0.5 A.
(a) What is the value of the current through the 4 Ω resistor? (1 mark)
(b) What is the value of the current through the 6 Ω resistor? (1 mark)
(c) Calculate the p.d., V, across the 6 Ω resistor. (Use the equation $V = IR$) (2 marks)
(d) What is the p.d. across the 4 Ω resistor? (1 mark)

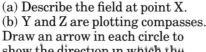

(Total: 5 marks)
(MEG, Syll. A, Paper 2)

88 The figure shows two circuits, (X) and (Y), which have been set up using similar components.

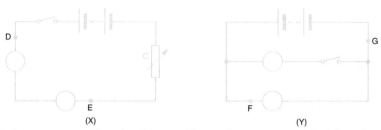

(X)    (Y)

(a) The circles represent identical lamps. Draw the correct symbol for a lamp. (1 mark)
(b) What do the following circuit symbols represent?

(i)      (ii)   (iii)

(3 marks)
(c) Which circuit, (X) or (Y), shows lamps connected in parallel? (1 mark)
(d) Add a meter to circuit (X) so that the potential difference across component C can be measured. (2 marks)
The switches are now closed.
(e) What difference is there, if any, in the current at points
(i) D and E,    (ii) F and G? (3 marks)
(f) (i) In which circuit will the lamps be brighter?
    (ii) In which circuit can one lamp be switched off and the other remain on? (2 marks)
(g) What effect does adjustment of component C in circuit (X) have on the lamps? (1 mark)
(Total: 13 marks)
(NEA, Syll. A, Level P)

89 The diagram shows a circuit containing three identical ammeters $A_1$, $A_2$, $A_3$.
State whether the following statements are true or false:
(i) $A_2$ gives the reading of the current through the cell;
(ii) the reading of $A_3$ is the current through the 6R resistor;
(iii) the reading of $A_1$ is less than the reading of $A_3$;
(iv) the reading of $A_3$ is equal to the reading of $A_2$ minus the reading of $A_1$.
(4 marks)
(WJEC, Paper 1)

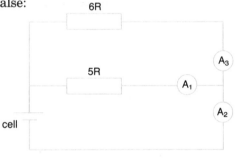

90 In a physics book a student reads that the resistance of a piece of wire depends upon
   its cross-sectional area
   its length
   the material of which it is made
The student decides to examine how the cross-sectional area of a piece of wire affects its resistance. He selects some pieces of resistance wire, each piece 1 m long.
(a) (i) Why should all the pieces of wire be made of the same material? (1 mark)
(ii) Why should they all be the same length? (1 mark)
(iii) In order to calculate the cross-sectional area the student needs to find the diameter of the wire. Name a suitable instrument for finding the diameter. (1 mark)
(iv) Why should the student find the diameter at several places along the wire? (1 mark)
(b) The circuit diagram shows the circuit which was used in this experiment. Using each of the wires in turn, the voltage was varied, and both the voltage and the current were measured.

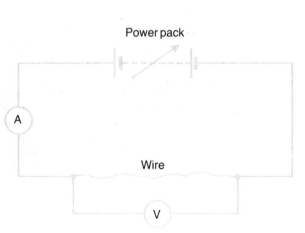

The results for three different wires are shown on the graph below.

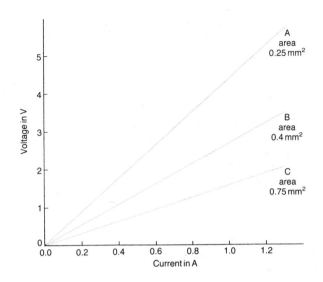

(i) For each wire find the resistance, showing how you do your calculation. (5 marks)
(ii) The cross-sectional areas of the wires are shown on the graph. What happens to the resistance of the wire as the cross-sectional area increases? (1 mark)
(Total: 10 marks)
(SEG, Syll. A, Paper 4)

91 This question is about the electrical system of a car.
Read the information below and then answer the questions which follow.
In a car there are two sources of electricity. A lead-acid battery is used to turn the starter motor to start the engine. Once the engine is running it turns the alternator (or dynamo). Usually the alternator provides sufficient electricity for all purposes when the engine is running, and the battery is then not used.
Some of the items which use electricity are listed in the table below.

| Item | Potential difference/V | Current/A |
| --- | --- | --- |
| starter motor | 12 | 200 |
| headlamp | 12 | 5 |
| tail/side lamp | 12 | 0.5 |
| heated rear window | 12 | 12 |
| heater fan | 12 | 6 |

(a) State the energy changes that occur in each of the following:
(i) the alternator (or dynamo)   (ii) the headlamps (4 marks)
(b) (i) Calculate the power developed in the starter motor when it turns the engine.
 (ii) If the starter motor operates for 5 seconds, how much energy is taken from the battery?
 (iii) Only part of the energy calculated in (ii) will be converted to *useful* work to turn the engine. What happens to the rest?
 (iv) The wires connecting the starter motor to the rest of the circuit are made from very thick copper wire. Why is this necessary? (7 marks)
(c) A driver keeps her car outside. She uses the car only for travelling to work, a distance of under three miles. After several months the battery 'goes flat'.
(i) Explain why this has happened.
(ii) Why is it more likely to happen during the winter? (2 marks)
(d) Why is it unwise to keep the heated rear window on all the time? (2 marks)
(Total: 15 marks)
(LEAG, Syll. B, Paper 2)

92 The following is a list of power ratings: 60 W; 250 W; 750 W; 1000 W; 3000 W.
Complete the table by selecting the power rating, from the above list, likely to be nearest to that of each appliance. (Each power rating may only be used once.)

| Appliance | Power rating |
|---|---|
| An electric blanket | |
| A high speed kettle | |
| A lamp | |
| A one-bar electric fire | |
| A vacuum cleaner | |

(2 marks)
(WJEC, Paper 1)

93 An electric light bulb is marked '240 V, 100 W'. It contains a length of fine tungsten wire about 1 metre long. The wire is wound in a coiled coil, as in the diagram. When switched on the wire reaches a temperature of about 2500°C.
(a) What is the power of this light bulb? (1 mark)
(b) When the bulb is on, the resistance of the wire is about 600 Ω. Why does the wire have to be as long as 1 metre? (1 mark)
(c) Why is the wire in a coiled coil? (1 mark)
(d) State **two** properties of tungsten which make it suitable for use in light bulbs. (2 marks)
(e) If you hold the back of your hand near a light bulb and switch the bulb on for less than 2 seconds your hand feels warm. If you then touch the bulb, the bulb is still cool. Why does your hand feel warm but the bulb is still cool? (2 marks)
(Total: 7 marks)
(MEG, Syll. A, Paper 2)

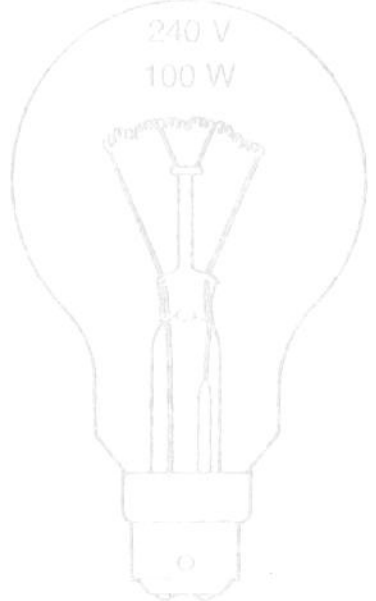

94 Electrical energy supplied to your home is measured in kilowatt hours.
(a) What is a kilowatt hour? (2 marks)
(b) The figures below show the readings on an electricity meter at the start and end of December.

kWh | 6 | 4 | 8 | 3 |
1000 100 10 1
1 December
**Figure 1**

kWh | 6 | 9 | 9 | 4 |
1000 100 10 1
31 December
**Figure 2**

How many kWh have been used during the month of December? (1 mark)
(c) Below is a list of appliances found in many homes, with their power ratings.

| Appliance | Power rating | Time used (h) | kWh |
|---|---|---|---|
| Television | 200 W | | |
| Electric kettle | 2 kW | | |
| Immersion heater | 3 kW | | |

A student arrives home in the evening, switches on the television at 7 p.m. and leaves it on until midnight. She makes three pots of tea using the kettle for five minutes on each occasion. She puts on the immersion heater for two hours, has a bath and goes to bed.
(i) Complete the table to show for each appliance
(A) the time for which it was used, (B) the number of kWh used. (3 marks)
(ii) What will be the total cost if 1 kWh costs 6p? (2 marks)
(Total: 8 marks)
(NEA, Syll. B, Level P)

95 The diagram (page 167) shows a radiant electric fire for use on a 240-V mains supply. It has four heating elements, each rated at 0.75 kW.
(a) Two important features of the fire are the shiny surface and the wide, heavy base. Explain the purpose of these. (3 marks)
(b) The four heating elements are controlled by three switches. One of these controls the middle **two** elements, the others control **one** element each.
(i) Complete the circuit diagram to show how this can be done. (3 marks)
(ii) What is the power rating of the fire? (1 mark)
(iii) The electricity board charges 7p for each kilowatt hour of energy. What is the cost of using the fire from 6 p.m. to 11 p.m.? (2 marks)

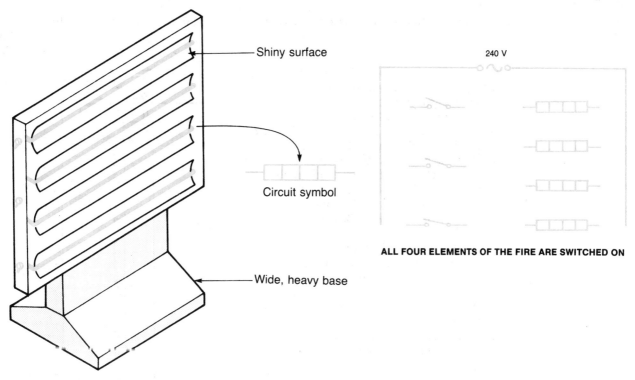

(iv) When used on the 240-V mains supply, what is the current drawn by the fire? (3 marks)
(v) What value fuse should be fitted to the plug? (1 mark)
(Total: 13 marks)
(LEAG, Syll. A, Paper 3)

96 Two lamps, one marked '240 V 100 W', the other marked '240 V 60 W', are used correctly in a house.
(a) Which lamp will be brighter? Give a reason for your answer. (1 mark)
(b) It costs 10p to run the 100 W lamp for a day. How much will it cost to run the other lamp for the same time? (1 mark)

The 100 W lamp is used in the kitchen and the 60 W lamp is used in a bedroom. Circuits A and B show two ways of connecting the lamps, with switches, to the mains supply.
(c) Give **two** reasons why circuit B is much better than circuit A.
(2 marks)
(Total: 4 marks)
(MEG, Nuffield, Paper 2)

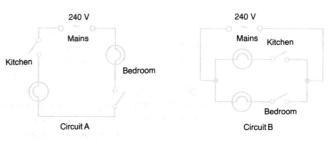

97 The figure shows a plug wired by a fifth-year pupil and connected to a hair dryer rated at 100 W, 240 V a.c.
(a) Several mistakes have been made. Name **four** of them. (4 marks)
(b) The earth pin on a plug is longer than the other two pins. Explain why. (2 marks)
(Total: 6 marks)
(NEA, Syll. B, Level P)

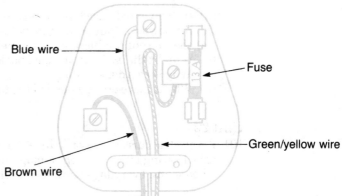

Plug viewed from the back

98 It is not wise to use an electric fire without a connection to the *Earth* pin. Why is this?
(2 marks)
(MEG, Nuffield, Paper 2)

99 In a factory which makes small electric motors one motor in every thousand is tested in the laboratory.

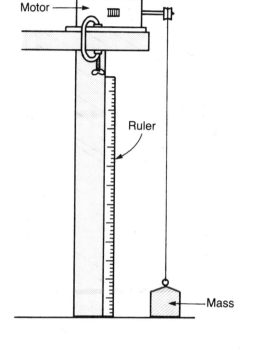

The motor is clamped to the edge of a bench-top as shown. A small mass is placed on the floor and connected by a string to the shaft of the motor. A ruler is fixed to the leg of the bench and the time taken for the motor to lift the mass between the 20 cm mark and the 70 cm mark is measured.

(a) (i) Why is the motor clamped to the bench? (1 mark)
 (ii) What method might be used to ensure that an accurate value is obtained for the time interval?
(2 marks)

(b) In one experiment the following results were obtained:

| Current through the motor | = 0.40 A |
| --- | --- |
| Voltage applied to the motor | = 2.0 V |
| Weight of load | = 5.0 N |
| Distance travelled | = 0.50 m |
| Time taken | = 10.0 s |

Calculate
(i) the work done in lifting the load; (2 marks)
(ii) the power output of the motor; (2 marks)
(iii) the power input of the motor; (2 marks)
(iv) the efficiency of the motor. (2 marks)
(Total: 11 marks)
(SEG, Syll. A, Paper 4)

100 Pumped-storage power stations are used to produce electricity during periods of peak demand. Water is stored in one reservoir and allowed to flow through a pipe to another reservoir at a lower level. The rush of water is used to turn turbines which are connected to generators.
(a) Why are pumped-storage power stations used to produce electricity in short bursts and not for the continuous generation of electricity? (1 mark)
(b) In one such power station 400 kg of water passes through the turbines every second after falling through a vertical height of 500 m. ($g = 10$ m/s$^2$).
Assuming that no energy is wasted, calculate
(i) the decrease in gravitational potential energy of 400 kg of water when it falls 500 m;
(3 marks)
(ii) the power delivered to the turbines by the water; (1 mark)
(iii) the power output of the generator; (1 mark)
(iv) the current produced by the generator if the output voltage is 20 000 V. (3 marks)
(c) At night the water is pumped back to the reservoir to be used again.
(i) Why is this done at night? (1 mark)
(ii) Why is more energy needed to pump the water back than is released when it falls?
(1 mark)
(Total: 11 marks)
(SEG, Syll. A, Paper 4)

**Unit 22**

101 An electric current may produce the following two effects:
(a) a heating effect, (b) a magnetic effect.
Choose **one** everyday example of each and say what it is used for. (4 marks)
(NEA, Syll. B, Level P)

102 The diagram shows an iron bar with a coil surrounding it. When a current flows in the direction of the arrows, the compass needles point as shown.

(a) What is the pole at A? (1 mark)
(b) By referring to the diagram at the beginning of the question, complete the following diagrams by showing the direction in which the compass needle points. (2 marks)

(Total: 3 marks)
(WJEC, Paper 1)

103 A coil of wire is connected in series with a battery, a rheostat and a switch.

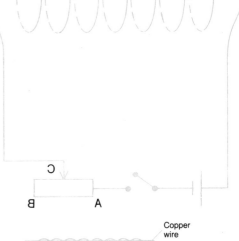

(a) Draw, on the diagram, the shape of the magnetic field **inside and outside** the coil when the switch is closed. (2 marks)
(b) A stronger magnetic field could be produced by winding the coil round a metal bar. (1 mark)
Name a suitable metal.
(c) If the slider, C, on the rheostat is moved towards B, what is the effect on
(i) the resistance of the circuit?
(ii) the current through the coil?
(iii) the magnetic field in the coil?
(3 marks)
(Total: 6 marks)
(WJEC, Paper 1)

104 The figure shows a solenoid. It consists of many turns of insulated copper wire. When an electric current is passed through the solenoid it behaves like a bar magnet.
(a) A piece of soft iron is placed inside the solenoid.
(i) What happens to the iron when
(A) the switch is closed,
(B) the switch is now opened?
(ii) Name a device which makes use of this behaviour. (3 marks)
(b) The soft iron is removed and replaced by a piece of steel. What happens to the steel when
(i) the switch is closed,   (ii) the switch is now opened? (2 marks)
(c) An alternating current is now passed through the solenoid.
(i) What is an alternating current?
(ii) What is the effect on the steel if it is slowly pulled through, and moved away from the solenoid? (3 marks)
(d) A cassette tape is made of plastic. There are many thousands of tiny particles on the surface of the plastic.
(i) From the list below choose **one** material suitable for making these particles.
    aluminium    cobalt    copper    zinc
(ii) Use the ideas developed above to explain how your cassette recorder removes the previous recording from the tape. (4 marks)
(Total: 12 marks)
(NEA, Syll. A, Level P)

105 The figure shows an electromagnet made by a pupil in the laboratory. The electromagnet is to pick up and release a metal object.

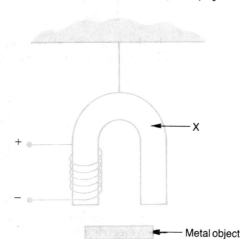

(a) Name a suitable material for part X. (1 mark)
(b) Why is it made from this material? (2 marks)
(c) The electromagnet will just lift a metal object of mass 0.15 kg. What will be the least force exerted by the magnet to do this? The strength of the gravitational field at the earth's surface, $g$, can be taken as 10 N/kg. (1 mark)

(d) Name a metal which the magnet will not attract. (1 mark)
(e) State **two** changes which the pupil could make so that a heavier metal object could be lifted by the electromagnet. (2 marks)
(Total: 7 marks)
(NEA, Syll. B, Level P)

106 (a)

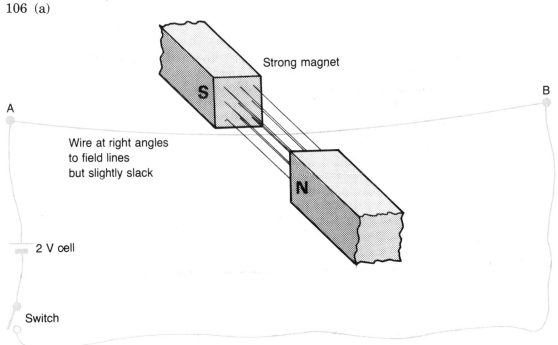

AB is a wire placed at right angles to the magnetic field lines of the magnet. The wire is supported so that the central portion is free to move up or down or sideways if a force acts on the wire.

The switch can be closed to demonstrate the *catapult field*.

What happens to the wire AB when:
(i) the switch is closed?
(ii) the single 2 V cell is replaced by several 2 V cells in series?
(iii) the connections to the cell(s) are reversed?
(iv) the 2 V cell is replaced by a transformer providing 2 V from the secondary and the switch is closed? (5 marks)
(v) The diagrams show the magnetic fields of:
(1) a wire carrying current, (2) opposite magnetic poles facing each other.

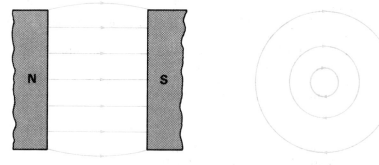

On the diagram below combine the fields (1) and (2) to give the resultant catapult field. Mark the direction in which the wire is catapulted. (3 marks)

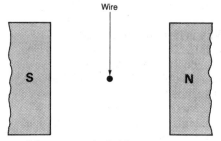

(b) A loudspeaker makes use of the catapult field in its operation. The cone moves to and fro when electric currents enter the coil.

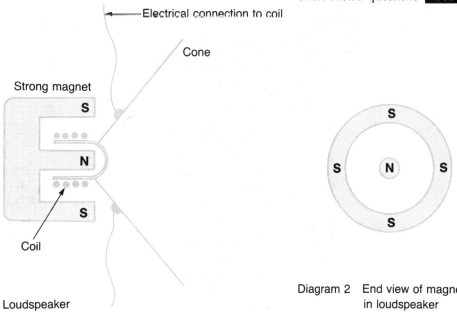

Diagram 1 Loudspeaker

Diagram 2 End view of magnet in loudspeaker

(i) Add the magnetic field lines to diagram 2, the 'end view of the magnet'. (2 marks)
(ii) Explain what causes the cone to move to and fro. (3 marks)
(iii) The demonstration in part (a) of this question uses a single wire. What is the advantage of using a coil of many turns in the loudspeaker? (2 marks)
(c) The loudspeaker is giving out a note of constant frequency 400 Hz.
A cathode ray oscilloscope is connected to the loudspeaker. The oscilloscope is adjusted so that the spot sweeps across the screen in 0.01 s.
Sketch what the trace on the screen of the oscilloscope will look like. (2 marks)
(Total: 17 marks)
(LEAG, Syll. B, Paper 3)

107 A student set up the electric motor shown in the diagram.
(a) The motor did not work. What was wrong with it?
(b) Why did this prevent the motor from working?
(c) When the mistake was corrected the motor turned.
   (i) Suggest **three** ways of making the coil turn faster.
   (ii) Suggest **two** ways of making the coil turn the other way.

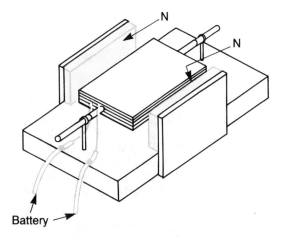

108 The diagram shows a model moving coil meter. When current flows through the coil the armature turns.
(a) What would be the effect of reversing the direction of the current flow?
(b) What would be the effect of reversing the direction of the current flow and also interchanging the magnets?
(c) Suggest two ways in which you might increase the angle through which the armature turns.
(d) What is the purpose of the springs?

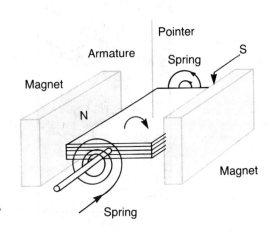

## Unit 23

109 The diagram shows a simple generator connected to a lamp. The lamp lights normally when the coil is rotating at its maximum speed.
(a) Describe what happens to the brightness of the lamp as the speed of the coil is increased from zero to its maximum speed.
(b) What would be the effect on the lamp of reducing the number of turns on the coil which is rotating at maximum speed?
(c) What would be the effect on the lamp of increasing the strength of the magnetic field, the coil having its original number of turns and rotating at maximum speed?

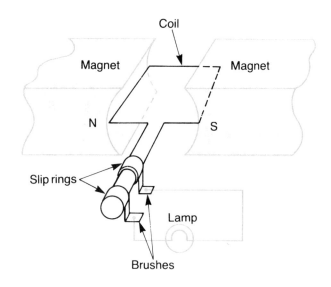

110 The diagram shows a circuit containing two coils of insulated wire wound on a soft iron core.

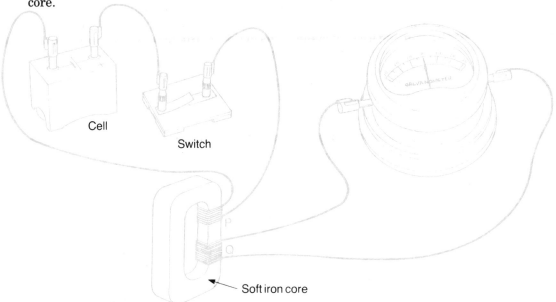

Coil P is connected to a switch and a cell; coil Q is connected to a galvanometer. As the switch is closed the galvanometer needle kicks to the right.
(a) What will happen to the galvanometer needle as the switch is **opened**? (1 mark)
(b) Explain why the galvanometer needle moves as the switch is closed. (2 marks)
(c) State **one** change that can be made so that the galvanometer needle kicks **further** to the right as the switch is closed. (2 marks)
(Total: 5 marks)
(MEG, Nuffield, Paper 2)

111

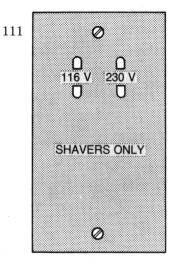

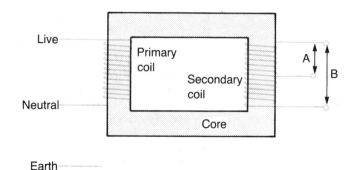

A bathroom razor socket contains a transformer. This is to isolate the user from the mains supply.
(a) On the right-hand diagram,
(i) add an S to show the correct position for a switch,
(ii) add an F to show the correct position for a fuse,
(iii) connect the earth lead to the correct point on the transformer. (4 marks)
(b) The transformer has two outputs, labelled A and B on the right-hand diagram. Complete the table below to show the number of turns and output voltage for each coil.
(5 marks)

| input voltage | primary turns | secondary turns | A or B | output voltage |
|---|---|---|---|---|
| 230 V | 5000 | 2500 | | |
| 230 V | 5000 | 5000 | | |

(Total: 9 marks)
(LEAG, Syll. A, Paper 2)

112 The diagram shows the beginning of the network of cables used to transmit electrical energy across the country.

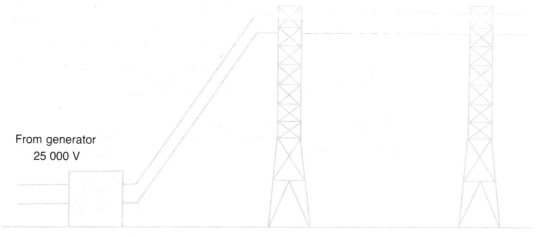

From generator
25 000 V

(a) What is the name of this network of cables? (1 mark)
(b) The turns ratio of the transformer is: $\dfrac{\text{number of secondary turns}}{\text{number of primary turns}} = \dfrac{11}{1}$.
Calculate the voltage of the transformer. (2 marks)
(c) Why is it called a step-up transformer? (1 mark)
(d) Why is electrical energy transmitted at high voltages? (1 mark)
(e) Why could it be dangerous to fly a kite near to high voltage overhead cables? (2 marks)
(f) Why is it safe for a bird to perch on a high voltage overhead cable? (1 mark)
(Total: 8 marks)
(MEG, Syll. A, Paper 2)

## Unit 24

113 The diagram shows parts of a cathode ray tube.
(a) Why is a high voltage needed between cathode and anode? (2 marks)

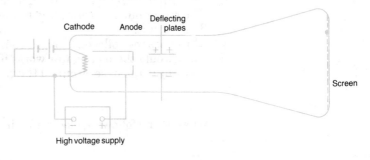

A spot is produced on the screen at **D**.
(b) On the diagram, draw the path of the cathode rays from the **anode** to **D**. (2 marks)
(c) The connections to the high voltage supply are changed over so that the anode is now **negative**. What happens to the spot? Give a reason for your answer. (2 marks)
(Total: 6 marks)
(MEG, Nuffield, Paper 2)

**Short answer questions**

114 A thermistor is a temperature sensitive resistor. Thermistors can be used to measure temperature. They are also used in temperature control applications.

(a)

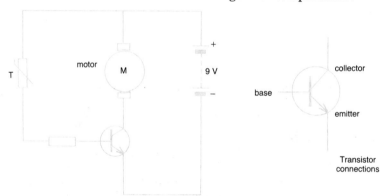

Some students used the apparatus above to investigate the electrical properties of a thermistor. The water bath in the left-hand diagram was used to vary the temperature. Their electrical circuit is shown in the right-hand diagram.

(i) Describe carefully how you would increase and measure the temperature in steps of about 20°C over the range 0°C to 100°C. (3 marks)

(ii) The students obtained the following results:

| Temperature/°C | 0 | 20 | 40 | 60 | 80 | 100 |
|---|---|---|---|---|---|---|
| Potential difference/V | 6.0 | 6.0 | 6.0 | 6.0 | 6.0 | 6.0 |
| Current/A | 0.08 | 0.16 | 0.28 | 0.62 | 1.18 | 2.40 |

Plot a graph of 'Current' against 'Temperature'. (4 marks)

(iii) Use your graph to find the current in the thermistor at a temperature of 50°C. (1 mark)
(iv) Calculate the resistance of the thermistor at 50°C. (2 marks)
(v) Calculate the resistance of the thermistor at 100°C. (2 marks)
(vi) How does the resistance of the thermistor change with temperature? (1 mark)

(b)

The circuit above was used by some students in a project to control an electric motor. They required the speed of the motor to change as the temperature changes. The thermistor, T, controls the base current in the transistor.

(i) What happens to the base current when the temperature increases?
(ii) Explain how the speed of the motor changes if the temperature increases.
(iii) The motor must be connected to the **collector circuit** of the transistor rather than the base circuit.
    How does the collector current compare with the base current?
(iv) Why would the motor, **not** work, if it were connected between the thermistor and the resistor in the base circuit of the transistor? (4 marks)

(Total: 17 marks)
(LEAG, Syll. B, Paper 2)

115 (a) What do the following electronic circuit symbols represent? (4 marks)

(b)

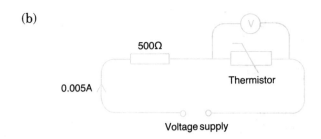

In the above circuit the voltmeter shows a reading of 4 V. Calculate the resistance of the thermistor. (2 marks)
(c) If the thermistor in the circuit in (b) is heated, what effect would this have on
(i) the resistance of the thermistor? (ii) the circuit current? (2 marks)
(d) Give two practical uses of thermistors in electronic circuits. (2 marks)
(Total: 10 marks)
(WJEC, Paper 1)

116 The circuit diagram shows how a transistor may be used to operate a switch which makes a light come on when it gets dark.
(a) What happens to the resistance of the light sensitive resistor when it gets dark?
(b) What effect does this change of resistance have on the collector current?
(c) Why is a variable resistor preferred to a fixed resistor at A?
(d) Why is it better to use a relay rather than to insert a bulb directly at B?
(e) What is the purpose of resistor $R_1$?

117 Complete the truth table for the system of logic gates shown.

| A | B | P | Q |
|---|---|---|---|
| 0 | 0 |   |   |
| 0 | 1 |   |   |
| 1 | 0 |   |   |
| 1 | 1 |   |   |

(2 marks)
(WJEC, Paper 1)

118 In the above circuit L acts like a switch which is **open** when the light falling on it is **dim** and is **closed** when the light is **bright**.
It now gets **dark**.
(a) What is the logic level of the input to the NOT gate? (1 mark)
The truth table for the NOT gate is shown below.

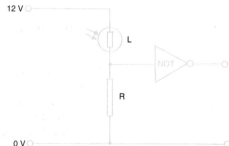

| Input | Output |
|-------|--------|
| 0 | 1 |
| 1 | 0 |

(b) What is the logic level of the output of the NOT gate? (1 mark)
(c) On the diagram, draw in a light emitting diode (LED) and a suitable resistor so that the LED will light when it is in the **dark**.
The symbol for an LED is (3 marks)

(d) Why is a suitable series resistor needed? (1 mark)
(e) Suggest a use for this type of circuit. (1 mark)
(Total: 7 marks)
(MEG, Syll. A, Paper 2)

119 (a)

The symbol shows a NAND gate connected as an inverter. Complete the truth table for it.

| Input | Output |
|-------|--------|
| 0 |   |
| 1 |   |

(1 mark)

(b) The two circuits A and B, include buzzers, an inverter and switches.
For each circuit say what happens when the switch is closed.
(2 marks)

(c) In a car, a buzzer sounds when a front seat passenger has **not** fastened his seat belt.
However the buzzer does not sound when there is no passenger.
Devise a circuit using **two switches**, an **inverter** and a **buzzer** which could be used. Draw your circuit in the space below. The inverter has already been drawn. (3 marks)

(Total: 6 marks)
(MEG, Nuffield Paper 2)

## Unit 25

120 The diagram represents a neutral atom.

(a) For this atom, fill in the table below.

| number of electrons | |
|---|---|
| atomic (proton) number | |
| number of neutrons | |
| nucleon (mass) number | |

(4 marks)

(b) The diagram represents an atom of lithium. In what way will an atom of another isotope of lithium differ from this one? (1 mark)

(Total: 5 marks)
(MEG, Syll. A, Paper 2)

121 This question is about radioactivity and its applications. When an alpha particle moves through a cloud chamber it leaves a straight, short, thick track.
  (a) (i)  What is an alpha particle? (1 mark)
      (ii) What do the tracks consist of? (2 marks)
      (iii) Why are the tracks short? (2 marks)
      (iv) Why are the tracks thick? (2 marks)
  (b) A radioactive tracer can be injected into the blood stream to allow the flow of the blood through the blood vessels to be monitored. Figure 1 shows the basic arrangement used. The graph in Figure 2 shows the results of monitoring the blood in the leg of a patient.

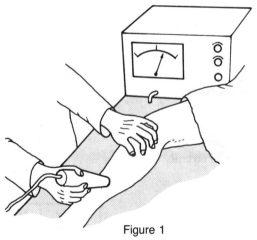

Figure 1

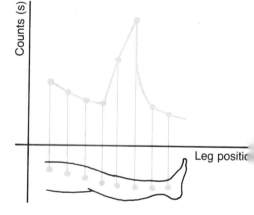

Figure 2

  (i)  Given that the particles from the tracer have to pass through the leg, what is the most likely type of particle to be emitted by the tracer? Explain your answer. (3 marks)
  (ii) Why does the graph show a peak above one part of the leg? (2 marks)

(iii) Why should the half-life of the tracer be quite short? (1 mark)
(iv) Why is it necessary for the radiographer operating the tracer detector to wear protective clothing? (1 mark)

(Total: 14 marks)
(NEA, Syll. B, Level P)

122 (a) Radioactive sources give off three types of ionizing radiation, alpha radiation, beta radiation and gamma radiation. The range of alpha radiation through air is only a few centimetres. Gamma radiation will pass through several centimetres of lead.

(i) You are provided with a GM tube and scaler. Describe in outline how a teacher would use this detector together with other necessary apparatus to compare the range of beta radiation through matter with the range of alpha and gamma radiations. You may add to the diagram if you wish. (4 marks)

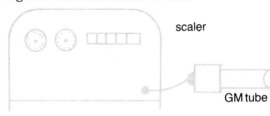

(ii) What results would you expect? (2 marks)
(iii) Explain the meaning of the term 'ionizing radiation'. (2 marks)

(b) This part of the question is about an application of radioactive sources.
Some plastic bottles are filled with a pain killing drug as they pass along a conveyor belt. It is important that all bottles contain exactly the same volume of the drug. This is done using two sensors at the end of the conveyor belt as shown in the diagram below. Each sensor faces a radioactive source. The sensors are connected to a system of logic gates. The conveyor belt is moving perpendicular to the paper.

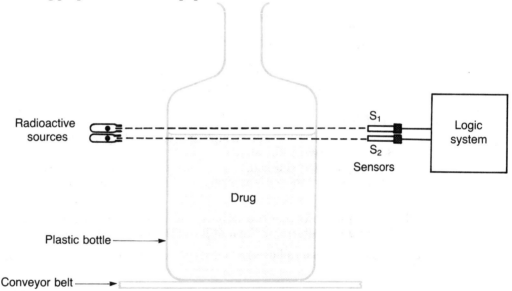

The bottle in the diagram is **correctly** filled. For such a bottle, sufficient radiation reaches the sensor, $S_1$, to cause a HIGH signal to enter the logic system. However, the output from $S_2$ into the logic system is LOW. Sensors $S_1$ and $S_2$ are designed so that their outputs can be only HIGH or LOW, nothing in between.

(i) Explain why the reading from $S_1$ is HIGH and that from $S_2$ is LOW. (2 marks)
(ii) Which type of radioactive source would be most suitable for this application, an alpha, beta or gamma source? Give reasons for your choice. (3 marks)
(iii) The logic system is made up from AND gates and NOT gates. Complete the 'truth tables' for these logic gates.

Truth Table

| Input | Output |
|---|---|
| LOW | HIGH |
| HIGH | |

Truth Table

| Input A | Input B | Output |
|---|---|---|
| LOW | LOW | |
| LOW | HIGH | LOW |
| HIGH | LOW | |
| HIGH | HIGH | |

(4 marks)

(iv) The logic system is shown below. Complete the truth table below for the logic system.

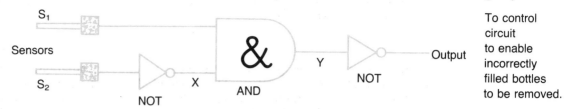

| State of bottle | Output from S | Output from S | X | Y | Output |
|---|---|---|---|---|---|
| under filled | HIGH | HIGH | LOW | LOW | HIGH |
| correctly filled | HIGH | LOW | | | LOW |
| over filled | | | HIGH | | |

(3 marks)
(Total: 20 marks)
(LEAG, Syll. B, Paper 2)

123 The diagram shows a stream of radiation, from a radioactive source, entering an electric field between two parallel plates P and Q.
When the electric field is on,
(i) which plate becomes positively charged?
(ii) name the radiation deflected along Z.
(iii) name the radiation deflected along X.
(iv) explain why radiation Y is undeflected.
(3 marks)
(WJEC, Paper 1)

124 The following is a list of particles:
   proton; neutron; electron; alpha particle; beta particle.
   State which **two** of the above
   (i) are positively charged,  (iii) occur in equal numbers in all neutral atoms,
   (ii) are found in the nucleus of an atom,  (iv) are identical.
   (4 marks)
   (WJEC, Paper 1)

125 A detector shows that the activity of a radioactive sample falls from 160 units to 20 units in 15 minutes. Determine the half-life of the sample, explaining your calculation.

126 The activity of a radioactive substance was measured using a counter.

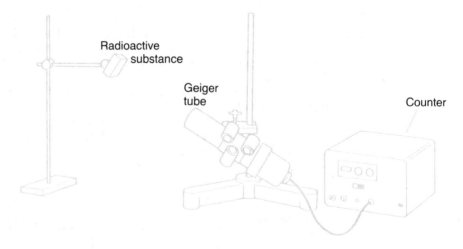

The graph shows how this activity varied with time. (The background count has already been deducted and can therefore be ignored.)
(a) Use the graph to find the activity after 10 hours. (Give your answer in *counts/minute*.)
(1 mark)
(b) Explain why the activity decreases with time. (2 marks)
(c) (i) Use the graph to find how long it takes, from the beginning of the experiment, for the activity to fall to 250 counts/minute.
(1 mark)

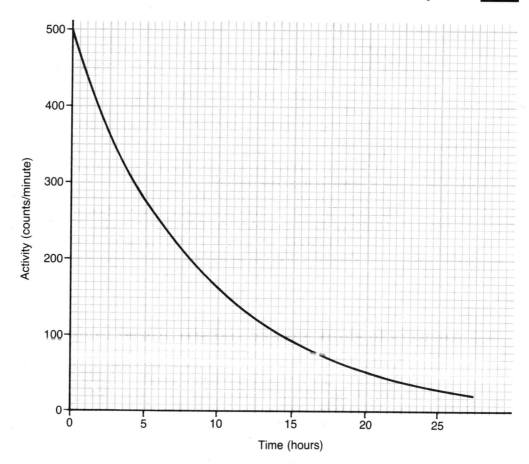

(ii) On the graph, sketch another curve, again starting at 500 counts/minute, for a different radioactive substance with a half-life *twice* as big. (2 marks)
(Total: 6 marks)
(MEG, Nuffield, Paper 2)

127 The symbol in the figure is on a cupboard door to warn of danger.
   (a) (i) What sort of materials would you expect to find inside the cupboard?
   (ii) Why are the materials dangerous? (2 marks)

(b) The figure below shows the container in which the material is stored. It is a solid wooden box with holes lined with metal.
   (i) Which metal is used for the lining?
   (ii) Why is this metal necessary? (2 marks)
   (c) Write down **two** safety precautions that should be taken when using the material in the capsule. (2 marks)
   (d) You would see the danger symbol
   (i) at which type of power station,
   (ii) in which department of a hospital? (2 marks)
   (Total: 8 marks)
   (NEA, Syll. A, Level P)

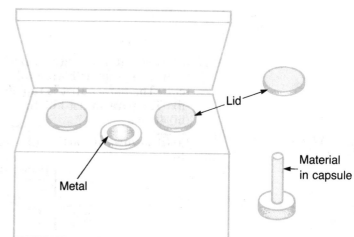

128 The reaction in a nuclear reactor can be represented simply by the equation
$$^{235}_{92}U + ^{1}_{0}n \longrightarrow ^{b}_{a}X + ^{d}_{c}Y + 3\,^{1}_{0}n + \text{Energy}$$
where X and Y are unspecified nuclides.

(a) What do the numbers 235 and 92 tell us about the particles in the uranium nucleus?
(2 marks)
(b) What is the name given to the process described by the equation? (1 mark)
(c) Complete the following sentence to describe the principle governing the energy production.
   The energy is produced by converting _____
   into _____. (2 marks)
(d) By using the reaction equation at the beginning of the question, complete the following equations:
(i) a + c =      (ii) b + d =      (2 marks)
(e) What property of the nuclides X and Y makes them a health hazard when the fuel rods are removed from the reactor? (1 mark)
(f) The neutron absorbed by the uranium is most likely to be a low energy (thermal) neutron, but the neutrons produced by the reaction are high energy neutrons which must be slowed down. Complete the following sentences:
   The neutrons are slowed down by a material called the _____.
   A suitable material would be _____.
(2 marks)
(Total: 10 marks)
(NEA, Syll. A, Level Q)

## Answers to short answer and structured questions

1 (a) The trolley moves down the runway with constant acceleration until the mass hits the floor. After this it travels at constant speed.
(b) (i) 10 cm   (ii) 100 cm/s or 1 m/s
(c) There is no resultant force on it.
(d) (i) increase   (ii) decrease   (iii) increase

2 (a) 30 m/s   (b) travelling at constant speed
(c) decelerating uniformly   (d) B
(e) The train slows down at a greater rate than it speeds up.
(f) 750 m

3 (a) Car A is travelling at uniform speed (20 m/s) for 30 s and then undergoes uniform deceleration.
(b) 40 s   (c) (i) 12.5 m/s   (ii) 3 m/s$^2$   (iii) 300 m
(d) The area under graph A is greater than that under graph B.

4 (a) 6 s   (b) 6 s   (c) 240 m

5 (a) 2 000 000 N
(b) Their propellers are put into reverse and the ships gradually slow down.

6 (a) Weight acts vertically downwards.
A buoyancy force acts vertically upwards.
(b) The two forces are equal and opposite.
(c) [graph of velocity vs time, inverted-U shape from V,N to F]

7 (a) (i) weight or gravity   (ii) friction due to air resistance   (iii) it increases
(b) (i) The graph is horizontal.   (ii) The two forces are equal and opposite.   (iii) 800 m
(c) (i) The parachutist is rapidly slowing down.
(ii) The parachutist hits the ground after 40 s and then slows down rapidly.
(iii) 40 m/s

8 (a) (i) downward, earth   (ii) weight   (iii) 5N   (iv) 0.5 kg
(b) (i) [diagram showing book with Friction due to air (small) upward and W downward]   (ii) 10 m/s$^2$

(iii)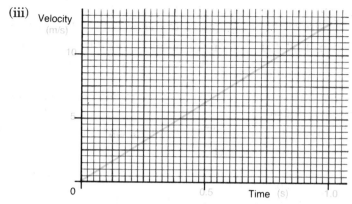

9 (a) 25 kg  (b) 42.5 N

10 (a)(i) The gravitational force between the two.  (ii) The tension in the string.
(iii) The frictional force between the tyre and the road.
(b) Inwards

11 6 m/s in the opposite direction to that in which the girl moves.

12 0.2 m/s

13 (a) potential  (b)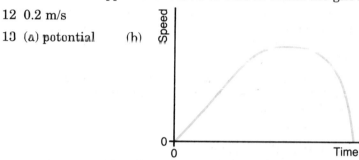

14 (a) 6.4 J  (b) 6.4 J  (c) 8 m/s  (d) 10 m/s$^2$  (e) 2 N

15 (a)(i) vertically downwards from the centre of crate C  (ii) 400 N  (iii) 1000 J
(iv) To place the crate on top of the others the truck must lift it slightly above the top crate.
(b) (i) potential  (ii) 1000 J  (iii) created, destroyed
(c) (i) kinetic  (ii) 1000 J  (iii) It turns into heat in the crate and floor.

16 (a) (i) The thinking distance is proportional to the speed.  (ii) 0.7 s
(b) (i) 89 m
(ii) 89 m so that if the first car stops dead, the second can pull up before it hits it.
(iii) about 18 m/s
(c) (i) 240 000 J  (ii) It turns into heat in the brake pads.  (iii) 30 m, 8000 N
(d) The kinetic energy possessed by the car is proportional to the square of its velocity. The constant braking force of the car is acting over a certain distance converts this to heat. The braking distance increases by a factor four.

17 (i) 200 J  (ii) 40 W

18 (a) The boat accelerates.  (b) They are equal.
(c) as heat in overcoming friction  (d) 240 W  (e) 86 400 J

19 (a)(i) 32 J  (ii) 32 J  (iii) 4 W  (b) (i) 24.5 J  (ii) 7 m/s

20 (a) A vector quantity has both size and direction e.g. a velocity of 3 m/s due N. A scalar quantity has only size e.g. a speed of 3 m/s or kinetic energy 5 J.

21 (a) 100 N  (b) (i) 6 m/s  (ii) 10 m/s
(iii) Mary is running at the same speed as the average speed of the world record holder.

22 (a) 20 N  (b) 27.5 N upwards

23 (a)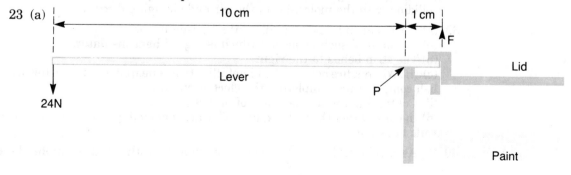

(b) 2.4 Nm or 240 Ncm   (c) 240 N   (d) 264 N
(e) Increase the size of the force applied to the end of the lever or increase the length.

24 (a) (i) 6000 J   (ii) 12 000 J   (iii) You can lift the load by pulling downwards.
(iv) You do not have to climb the ladders carrying the load.
(b) 200 W   (c) (i) 100 N   (ii) ¾ or 75%

25 (a) (i) measuring   (ii) water   (iii) 70 cm$^3$   (iv) level with the 90 cm$^3$ mark
(b) (i) balance   (ii) 8 g/cm$^3$

26 (a) (i) 30 cm$^3$   (ii) 27 cm$^3$   (b) (i) 27 g   (ii) 0.9 g/cm$^3$   (c) less than 0.9 g/cm$^3$

27 (a) 24 cm$^2$   (b) 8 cm$^3$   (c) 64 g

28 (a) Measure the length of the scale and the average width and multiply the two together.
(b) 3000 N   (c) 375 000 N/m$^2$
(d) (i) 150 m   (ii) 120 000 J   (iii) 12 000 W or 12 kW

29 (a) 100 000 N   (b) 100 000 N   (c) 1000 Pa (N/m$^2$)   (d) 25 000 N

30 The pressure depends on three factors: the density of the liquid, the depth of the liquid and the gravitational field strength.

31 (a) The same pressure is above each piston.
(b) (i) It is ten times greater.
(ii) Pressure = Force/Area, so the forces are in the same ratio as the areas.
(c) Oil is almost incompressible, whereas air is easily compressed.
(d) The change in the volume of oil in the two pistons must be the same. Thus the rise or fall will be inversely proportional to the ratio of the areas.

32 (a) 0–20 N: In this range equal increases in load produce equal amounts of compression.
(b) Springs in mattresses compress when people lie on them.
(c) The turns are then touching each other.   (d) 6.5 cm.

33 (b) (i) 15.4 cm   (ii) 12 cm   (iii) 1.25 N

34 (a) The springs represent the forces between molecules.
(b) When a solid is heated the molecules vibrate about their average position with greater amplitude. The spheres in the model should be made to vibrate with greater amplitude.
(c) In a liquid the molecules move throughout the liquid. In a solid molecules stay in the same average position.
(d) Energy is needed to separate the molecules.

35 (i) solid   (ii) gas   (iii) liquid

36 (a) The smoke particles reflect some of the light and can thus be seen.
(b) They have small enough mass to move slightly when buffeted by molecules, but are large enough to be seen under a microscope.
(c) The smoke particles 'jiggling' about in a random way.
(d) They are very small and moving very fast.
(e) None.

37 (a) Brownian motion
(b) (i) The brown colour of the bromine will have moved a very small distance up the tube.
(ii) The brown colour will have filled most of the tube, but the brown colour will be less strong as one goes up the tube.
(c) (i) 2000 m   (ii) See Fig 8.4.
(iii) The air should be removed from the tube before the bromine is let in.
(d) At a particular temperature some bromine molecules have much more energy than the average for all of them. These more energetic molecules have enough energy to leave the liquid. The average energy of those left behind decreases, therefore the liquid is cooler.

38 (a) Some of the molecules with much more than the average kinetic energy leave the liquid and move through the air colliding with its molecules. Eventually they will spread out to some distance from the spillage.
(b) The spreading is slow because the molecules which escape from the scent are continually colliding with the molecules in the air and changing direction.

39 (a)(i) The bar must be heated.   (ii) a bunsen burner
(iii) A material such as plastic which is a good heat insulator.
(b) The strip bends to the right.
(c) (i) When a fire occurs, the bimetallic strip is heated and bends towards and touches the bell contact, thus completing the electric circuit.
(ii) Controlling the temperature of an oven.
(d) In hot weather the rails expand. This arrangement gives room for expansion without the rails buckling.

40 (a) As a solid is heated the molecules vibrate with greater amplitude and the average

distance between neighbouring molecules increases slightly. Thus the solid becomes larger.
(b) When the rivets cool they contract and the two plates are squeezed together between the ends of each rivet.
(c)

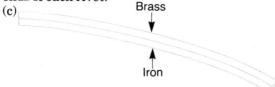

41 (a) (i) temperature   (ii) mercury   (iii) alcohol
(b) (i) 100°C   (ii) 373 K
(c) (i) The clinical thermometer is arranged to give a large increase in the length of thread for a relatively small temperature change. It is thus more accurate.
   The glass tube containing the mercury has a very narrow section in it. This prevents the mercury from returning to the bulb when the thermometer is removed from the patient. It can thus be read easily.
(ii) 310 K
(d) a pyrometer

42 There are more air molecules in the tyre to collide with the walls each second.

43 (a) 2083, 3125, 4348.
(b) This graph is a straight line through the origin and points (60, 2500) and (120, 5000).
(c) P is proportional to 1/V. That is, PV = a constant.
(d) 0.000 25 m³

44 (a) 1.5 litres   (b) 0.67 atmospheres

45 (a) Absolute zero is the lowest attainable temperature.
(b) Temperature is a measure of the kinetic energy of a substance. At absolute zero all motion ceases and thus a substance at this temperature has no kinetic energy. Absolute zero is thus a better zero than 0°C at which temperature substances do possess kinetic energy.

46 (a) (i) The molecules move faster.
   (ii) The pressure increases: Moving faster, the molecules collide with the walls more violently and more often.
   (iii) The molecules inside the can are moving faster and making more violent collisions with the lid than those on the outside. The pressure inside is greater than that outside, so the lid lifts off.
(b) The evaporation and eventual boiling of the water means that the density of molecules in the can is much higher than without the water. Molecules of water vapour as well as air collide with the lid increasing the pressure even further.
(c) If an aerosol can is heated the extra pressure can be enough to make the can burst.

47 (i) less than 30°C   (ii) greater than 30°C.

48 (a) 22 500 J/s   (b) and (c)

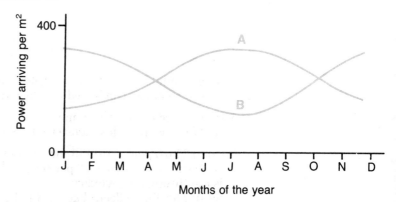

(d) July or August. In summer less space heating is required and so less energy is extracted from the tanks. The panels are also receiving energy at a faster rate.
(e) $5.25 \times 10^9$ J   (f) 2.7 days
(g) This answer assumes 100 per cent efficiency in the panels and heat exchangers.

49 (a) It comes from the kinetic energy of the water molecules which are no longer free to move about but only to vibrate.
(b) 167 000 J

50 (a) As steam condenses it releases all its latent heat into the skin.
(b) Evaporation takes places from the wet cloth. The energy needed for this evaporation is taken from the milk which thus cools.

51 (a) When water is converted to steam, energy is needed to increase the separation of the molecules against the force of attraction between them.
(b) 0.24 kg

52 (a) See the answer to question 37(d).
(b) heating; blowing air across the surface.
(c) Evaporation takes place from the surface of a liquid at all temperatures, whereas boiling only occurs over a certain temperature and occurs throughout the liquid.

53 (a) conduction   (b) convection   (c) radiation

54 (a) (i)

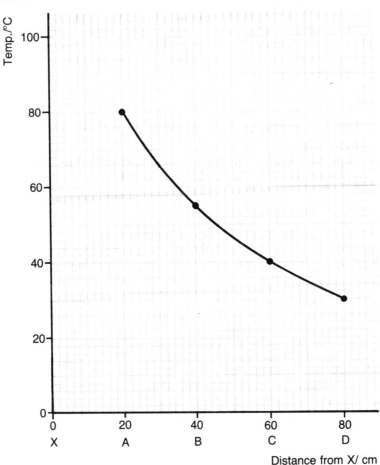

(ii) 66°C

(b) (i) conduction   (ii) two of: conduction, radiation, convection
(iii) Heat loss can be reduced by wrapping the bar in an insulating material (lagging).

55 (a) Upwards above the flame.
(b) Warm water is less dense than cold water and thus rises.

56 (a) (i) 35 per cent   (ii) Fill the cavity with an insulating foam.
(b) Insulating the roof: 25 per cent of the energy is lost via the roof, whereas only 10 per cent is lost through the windows.
(c) (i) The foil reflects back into the room radiant energy which would otherwise pass through the walls.
(ii) This prevents heat being lost by hot air rising up the chimneys.

57 (a) £74.86   (b)(i) loft   (ii) double glazing
(c) floor insulation, draught exclusion
(d) (i) £5000 million   (ii) 10 years
(e) (i) If the air gap is wide convection currents in it will increase.
(ii) If the glass is very thick the window will be less clear.

58 (a) (i) Less, because the temperature gradient of the thicker glass is less.
(ii) There is a thickness of still air (which is a good insulator), trapped between the layers of glass.
(iii) It reduces the noise entering a building.

(b) (i)

(ii) The air cannot circulate in the insulated cavity. There are thus no convection currents and heat loss is reduced.

59 (a) The disturbance which causes the wave is in the same direction as the wave travels.
   (b) Transverse: water, radio, light.   Longitudinal: sound

60 (a) (i) 4 cm   (ii) 40 cm   (b) 3.2 m/s

61 (a) (i) They travel with the same speed in all directions.
   (ii) 10 Hz   (iii) 2 cm   (iv) 20 cm/s   (v) They are reflected.
   (vi)

62 (a)   (b) (i) 12 mm   (ii) 8 Hz

63 (a) The waves will spread out and become almost circular within the harbour. The distance between waves will remain the same.
   (b) diffraction
   (c) The wavelength is shorter.
   (d) The diffraction (spreading out) would be less.

64 (a)   (b) (i)

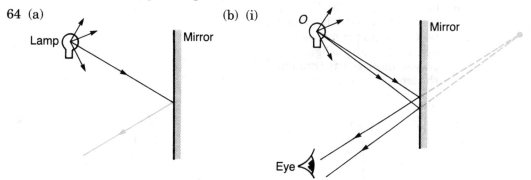

(ii) The image cannot be placed on a screen as rays do not actually cross here.

(c)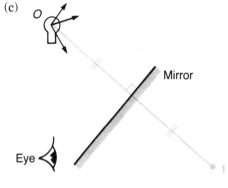

(a), (b) and (c) see diagram.
(d) It gives a narrow angle of view.

66 (a)(i) refraction    (ii) critical angle
(b)(i) The speed of light increases as it goes from glass to air.
(ii) The ray hits the surface at more than the critical angle and is reflected.

67 (a) X is the angle of incidence; Y is the angle of refraction.
(b) (i) Z is the angle of incidence.    (ii) (A) total internal reflection
(B) The incident angle is too large for refraction to take place.
(C) In reflectors.
(c) The transmission of telephone conversations by passing light along fibre optics.

68 (a) The ray is refracted towards the normal.
(b) The ray weakly reflected within the fibre but mostly refracted outside.
(c)
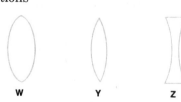
(d) to transmit telephone conversations

69 (a) W and Y    (b) Z    (c)
(d) a parallel sided block of glass
(e) Glass and perspex are
denser than air, so light
travels more slowly through
the lenses than air.

70 (a)
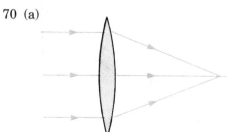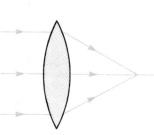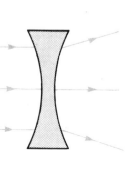

(b) (i) 1 cm below *I*

(ii)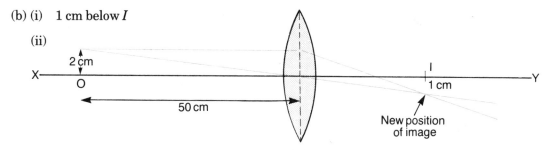

(iii) The image moves closer to the lens and gets smaller.

(c)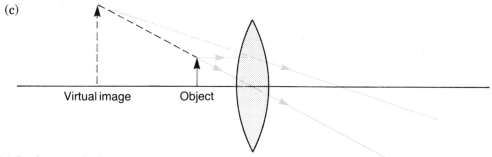

71 (a) Light travels faster than sound.
   (b) On entering glass, blue light slows down more than red light.
   (c) The rays continue to diverge until they reach the lens, then they are bent inward and come to one point at the centre of the film.

72 (a) long sight    (b) away from him
   (c) converging    (d)

73 thin or weak, behind, converging, on

74 long, in front of, diverging, on

75 (a) The spreading could be increased by making the slit narrower.
   (b) The wavelength remains the same.
   (c) The diffraction (spreading out) would be greater.

76 (a) The wave nature of light (diffraction, interference).
   (b) (i) bright   (ii) dark bands    (c) (i) bright   (ii) bright   (iii) dark
   (d) The fringe separation would become less (fringes closer together).
   (e) The fringe separation would become greater (fringes further apart).

77 (a) Longest: radio.   Shortest: ultra-violet.
   (b) 3 000 000 Hz
   (c) (i)  To prevent microwaves getting out of the oven and irradiating people.
       (ii) In a microwave oven only the food is heated. In a conventional oven energy is wasted in heating the air surrounding the food.
       (iii) heat
       (iv) (A) The wavelength is smaller.   (B) The frequency remains the same.

78 (a) (i) B, C and D    (ii) B, C and D    (iii) (A)
   (b) (i) refraction   (ii) diffraction   (iii) reflection

79 (a) Light can be seen by the human eye, infrared radiation cannot.
   Light has a higher frequency and a smaller wavelength than infrared radiation.
   (b) Thermal imaging cameras which can 'see' in the dark. These cameras can detect the presence of people or objects hidden under debris, for example.

80 750 m

81 (a) By reflection off the sea bed or an object (e.g. fish).
   (b) 750 cm    (c) 375 m

82 (a)

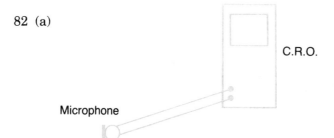

(b) (i) The same frequency (note), but louder.
(ii) A higher frequency (note), but the same volume.

83 (a) The S pole of magnet $Q$ is the furthest one from magnet $P$.
(b) They will repel.   (c) The region in which magnetic forces are experienced.

84 (a) There is no field at $X$. It is a neutral point.
(b)

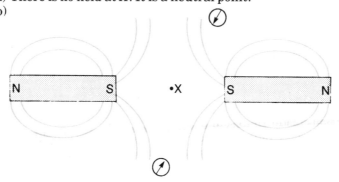

(c) They will move further apart.
(d)

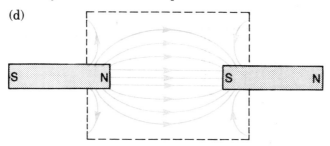

85 (a) By contact or by induction.   (b) Different

86 (a) The leaf will rise.   (b) The leaf will fall.   (c) The leaf will fall.

87 (a) 0.5 A   (b) 0.5 A   (c) 3 V   (d) 2 V

88 (a)

(b) (i) switch   (ii) cells   (iii) variable resistor   (c) Y
(d) The voltmeter is added in parallel with (across) C.
(e) (i) none   (ii) The current at F is half that at G.
(f) (i) Y   (ii) Y
(g) Increasing the value (resistance) of C will dim both lamps and vice-versa.

89 (i) true   (ii) true   (iii) false   (iv) true

90 (a) (i) Pieces of wire made from different materials will have different resistances even if they are otherwise identical.
(ii) The length of wire also affects the value of the resistance.
(iii) calipers or micrometer
(iv) To obtain an average value as the wire may not be uniformly thick.
(b) (i) A 4.4 Ω;   B 2.8 Ω;   C 1.7 Ω
(ii) As cross-sectional area increases the resistance decreases.

91 (a) (i) kinetic to electrical   (ii) electrical to light
(b) (i) 2400 W   (ii) 12 000 J   (iii) It is converted to heat in the connected wires.
(iv) They can carry a very large current.
(c) (i) She uses the starter motor frequently but does not drive far enough to recharge the battery sufficiently.
(ii) In winter the oil is thicker and the starter motor needs more current to turn the engine. It is also likely to take longer to start the engine.
(d) It draws current from the battery and may 'flatten' it.

Short answer questions: answers  189

92  Electric blanket 250 W; high speed kettle 3000 W; lamp 60 W; one-bar electric fire 1000 W; vacuum cleaner 750 W.

93  (a) 100 W    (b) To have a large enough resistance.
    (c) So that the coil will fit inside the bulb.
    (d) Tungsten is not brittle and it has a high melting point.
    (e) The radiant energy passes through the glass but is absorbed by the hand.

94  (a) The energy used by a 1 kW electrical appliance running for one hour.
    (b) 511 kWh
    (c) (i)

| Appliance | Power rating | Time used (hours) | kWh |
|---|---|---|---|
| Television | 200 W | 5 | 1 |
| Electric kettle | 2 kW | 0.25 | 0.5 |
| Immersion heater | 3 kW | 2 | 6 |

    (ii) 45 p

95  (a) (i) To reflect radiant heat into the room.
        (ii) To make it very difficult to accidentally tip over the fire.
    (b) (i)

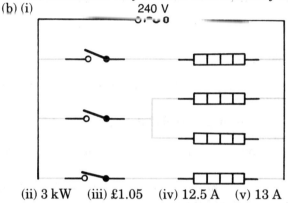

    (ii) 3 kW    (iii) £1.05    (iv) 12.5 A    (v) 13 A

96  (a) The '240 V, 100 W' lamp, because it gives out the most energy in each second (100 J).
    (b) 6 p
    (c) In circuit A both lamps will be dim as they share the 240 V supply, whereas in circuit B they are each connected to 240 V supply.
    In circuit A either both lamps are on or both are off, whereas in circuit B they can be switched on and off independently.

97  (a) Four of the following:
    1 The brown wire is connected to the neutral terminal. It should be connected to the live one.
    2 The blue wire is connected to earth. It should be connected to the neutral terminal.
    3 The green/yellow wire is connected to the live terminal. It should be connected to earth.
    4 The outside cable insulation has been cut back too far, so that the cable is not properly clamped by the clench and all the strain comes on the terminals.
    5 There is a sharp kink in the green/yellow wire.
    (b) When the plug is pushed into a socket, the longer earth pin opens gates to the live and neutral connections.

98  The earth connection ensures that the case of the fire is always at zero volts. Without it, the case could become live (possibly due to the live wire touching it) and anyone touching the case would be electrocuted.

99  (a) (i)   To keep the motor from moving and also to prevent some of the electrical energy being converted to vibrational energy
        (ii)  Using a stopwatch or a pair of photodiodes connected to an electronic timer.
    (b) (i) 2.5 J    (ii) 0.25 W    (iii) 0.8 W    (iv) 0.3125 or 31.25%

100 (a) In continuous operation the upper reservoir would soon empty.
    (b) (i) 2 000 000 J or 2 MJ    (ii) 2 MW    (iii) 2 MW    (iv) 100 A
    (c) (i) There is spare generating capacity at night as the demand for electricity is lower than in daytime.
        (ii) This is due to the inefficiency of the pumps.

101 (a) A heating coil in an electric fire.
    (b) Making an electromagnet or relay.

102 (a) N  (b) (i)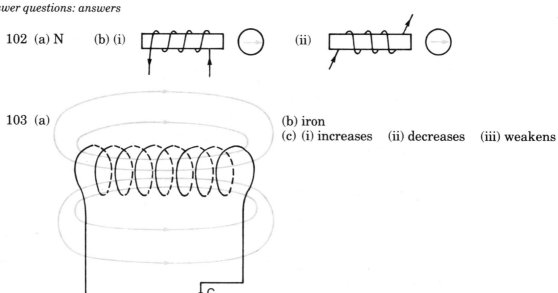

103 (a)  (b) iron
(c) (i) increases  (ii) decreases  (iii) weakens

104 (a) (i) (A) The iron is magnetized. (B) The iron loses its magnetism.
(ii) an electromagnet
(b) (i) The steel is magnetized.  (ii) The steel retains its magnetism.
(c) (i) A current whose direction of flow reverses at regular intervals.
(ii) The steel is demagnetized.
(d) (i) aluminium
(ii) An alternating current is passed through the erase head. This returns the magnetic particles to a random arrangement.

105 (a) soft iron  (b) It is easily magnetized and demagnetized.
(c) 1.5 N  (d) copper, zinc
(e) Increase the current and/or increase the number of turns of wire carrying the same current.

106 (a) (i) The wire moves up.  (ii) The wire moves up further  (iii) The wire moves down.
(iv) The wire will vibrate up and down.
(v)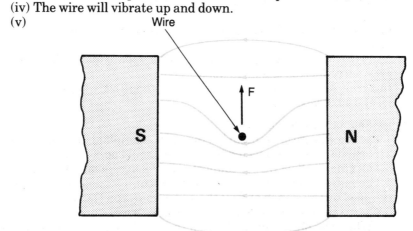

(b) (i)  (ii) The changing current in the coil leads to a changing magnetic field and hence to a changing force on the cone. The cone thus vibrates.
(iii) The force is larger.

(c) (4 complete waves)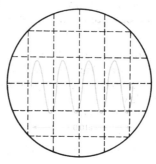

107 (a) The two magnets both have the same pole facing inwards.
(b) This means that there is a very weak magnetic field between the magnets, and therefore very little force on the sides of the coil.
(c) (i) Stronger magnets, more turns on the coil, a larger current.
(ii) Interchange the two magnets, reverse the direction of current flow.

108 (a) The direction of rotation of the coil would reverse.
(b) The coil would continue to rotate in its original direction.
(c) Increase the strength of the magnets; reduce the strength of the springs; increase the current.
(d) To control how far the armature turns for a given current, and to bring the coil back to its original position when the current is turned off.

109 (a) The lamp will be out or very dim for slow speeds. As the speed increases the brightness of the lamp will increase until it reaches its normal brightness when the coil is rotating at its maximum speed.
(b) The lamp would become less bright.
(c) The lamp would become brighter.

110 (a) The needle will flick one way and return.
(b) As the switch is closed current starts to flow one way through P. This produces a magnetic field in the core. The growing magnetic field in the core induces a voltage across Q and a current flows in Q while the magnetic field is increasing.
(c) Use more cells to give a bigger current.

111 (a)

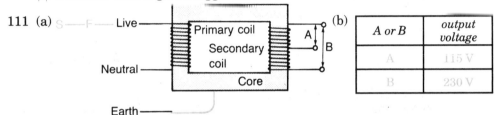

112 (a) The National Grid  (b) 265 000 V  (c) The voltage is increased.
(d) To reduce the current flowing and hence the energy lost due to heating the cables.
(e) The kite might touch the cables and thus connect the person flying it to the high voltage.
(f) The whole of the bird is on the wire, i.e. at only one voltage. There is no potential difference across the bird.

113 (a) To accelerate the electrons.
(b) The negatively charged electrons are attracted upwards by the positively charged plate.

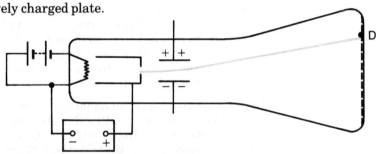

(c) The anode now repels all the electrons back to the cathode so the spot is no longer seen.

114 (a) (i) The water bath is heated by standing the beaker on a tripod and gauze. Once the bunsen has heated the water to about the required temperature it is removed, the water well stirred and the temperature of the water taken with the thermometer as near to the thermistor as possible.
(iii) 0.42 A  (iv) about 14.3 Ω  (v) 2.5 Ω  (vi) decreases
(b) (i) It increases.
(ii) The base current increases and hence the collector current also increases and the motor speeds up.
(iii) It is greater.
(iv) The current would be too small to drive the motor.

115 (a) (i) capacitor  (ii) transistor  (iii) light dependent resistor (l.d.r.)  (iv) fuse
(b) 800 Ω  (c) (i) The resistance would decrease.  (ii) The current would increase.
(d) As resistance thermometers, particularly at low temperatures.
To safeguard against current surges when circuits are switched on.

116 (a) The resistance increases.
(b) The collector current increases.

(c) The variable resistor is used to alter the potential of point $A$. If the potential of $A$ increases, the base current increases. This then increases the collector current.
(d) The relay requires less current to operate it than the bulb does to make it bright.
(e) Resistor $R_1$ limits the current through the base of the transistor and prevents damage.

117

| A | B | P | Q |
|---|---|---|---|
| 0 | 0 | 1 | 1 |
| 0 | 1 | 1 | 0 |
| 1 | 0 | 0 | 1 |
| 1 | 1 | 0 | 1 |

118 (a) zero  (b) one  (c)
(d) To restrict the maximum current that can pass through the l.e.d. thus saving it from damage.
(e) To switch on an indicator light at night.

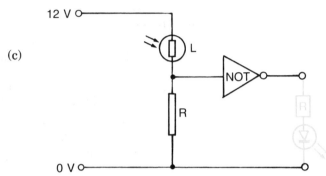

119 (a)

| Input | Output |
|---|---|
| 0 | 1 |
| 1 | 0 |

(c)

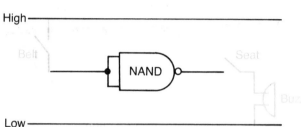

(b) **A** The buzzer sounds.
**B** Nothing.

(c) If NAND inputs float LOW, otherwise, reverse power lines for both belt and buzzer.

120 (a) 3 electrons; 3 protons; 4 neutrons; mass number 7.
(b) It will have a different number of neutrons and a different mass number.

121 (a) (i) A helium nucleus, $He^{++}$.  (ii) vapour droplets
(iii) Alpha particles have a range in air of a few centimetres.
(iv) Alpha particles cause many ion pairs per centimetre of track.
(b) (i) Beta particles.
(ii) The flow of blood is restricted here thus leading to a build up of radioactive material.
(iii) So that it does not irradiate the body for a long time.
(iv) So that the radioactive tracer does not get onto the body or clothes of the operator. The operator is working with the material all the time so could absorb a dangerous level of radiation.

122 (a) (i) and (ii) Set up an alpha source just in front of the GM tube and record the count rate. Place first one then two sheets of paper between the source and the tube and note that the count rate drops markedly. Repeat the experiment with a beta source a centimetre or two away from the GM tube and this time place sheets of aluminium between the two. It takes a few millimetres of aluminium to reduce the count rate significantly. Repeat the experiment a third time using lead between the gamma source and the tube. A few centimetres of lead are needed to reduce the count rate noticeably.
(iii) Radiation which causes the atoms of the material through which it passes to split into positive and negative ions.
(b) (i) There is no liquid between $S_1$ and the source to stop radiation, whereas radiation is stopped by the liquid before it reaches $S_2$.
(ii) Beta radiation has a range of a few millimetres in a liquid. Beta radiation would therefore not reach $S_2$.
(iii) NOT truth table

| input | output |
|---|---|
| LOW | HIGH |
| HIGH | LOW |

AND truth table

| input A | input B | input |
|---|---|---|
| LOW | LOW | LOW |
| LOW | HIGH | LOW |
| HIGH | LOW | LOW |
| HIGH | HIGH | HIGH |

(iv)

| State of bottle | Output from $S_1$ | Output from $S_2$ | X | Y | Output |
|---|---|---|---|---|---|
| under filled | HIGH | HIGH | LOW | LOW | HIGH |
| correctly filled | HIGH | LOW | HIGH | HIGH | LOW |
| over filled | LOW | LOW | HIGH | LOW | HIGH |

123 (i) Q  (ii) beta  (iii) alpha
 (iv) It consists of gamma rays which are not charged.

124 (i) proton, alpha particle
 (ii) proton, neutron
 (iii) proton, electron
 (iv) electron, beta particle

125 5 minutes. The activity falls to one-eighth of its original value. Thus 3 half-lives have elapsed in 15 minutes.

126 160 counts/minute
 (b) Atoms are decaying all the time. As time passes there are fewer and fewer left to decay. Thus the count rate falls.
 (c) (i) 6 hours
 (ii)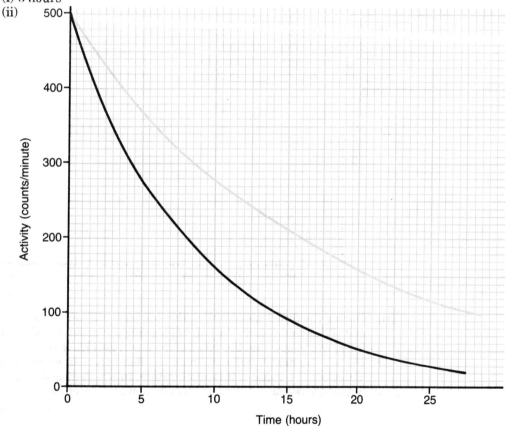

127 (a) (i) Radioactive sources.
  (ii) They give out radiation which can damage the tissues of the human body.
 (b) (i) lead
  (ii) It is very dense and is the most effective material for stopping radiation.
 (c) Hold the source in tongs, not the hand; point the source away from the body, never towards it.
 (d) (i) nuclear  (ii) X-ray (radiography) and radiology/radiotherapy.

128 (a) 235 is the total number of protons and neutrons in each uranium nucleus; 92 is the number of protons in each nucleus.
 (b) fission  (c) mass, energy  (d) a + c = 92, b + d = 233.
 (e) They are radioactive.  (f) moderator, graphite

## Longer questions

The questions are numbered according to the relevant unit in the text. However, some questions involve work covered in more than one unit, and in these cases the questions occur in the last relevant unit. Such questions thus provide an opportunity for revision of work in an earlier unit.

Each question is followed by a few hints about how to answer it (printed in green type). These hints are NOT intended as 'model answers'; they are merely meant to indicate some of the important physical principles involved in the question. The hints often refer back to the text and it is hoped that they will aid your understanding of the topic.

Answers to the numerical parts of questions are given after the last question (see page 215).

### Unit 2

1 (a) Describe an experiment to measure $g$, the acceleration of free fall, by using a freely falling object. Your answer should include
(i) a labelled diagram of the apparatus used, (4 marks)
(ii) a list of the measurements that must be made, (2 marks)
(iii) a note of **two** precautions that could be taken to ensure reliable measurements and a reasonable result, (2 marks)
(iv) an indication of how $g$ is calculated from the measurements. (2 marks)

(b) A flea jumps **vertically** into the air. The first drawing shows it pushing against the ground and the second one shows it in the air.

The simplified graph below shows how its speed varies with time starting from when it prepares to jump until it reaches the top of its jump.

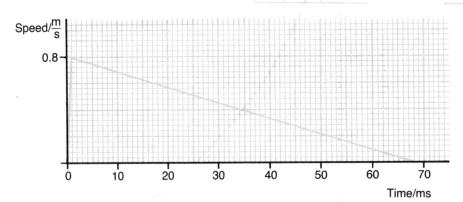

(i) During the first 1 ms (0.001 s), the flea pushes against the ground and accelerates. At the end of this time it is on the point of leaving the ground. If the acceleration during this time is 800 m/s$^2$, show that the speed of the insect as it leaves the ground is 0.8 m/s. (2 marks)
(ii) During the next 0.067 s (67 ms), the flea moves upwards with a constant acceleration. Calculate the value of this acceleration. In which direction is it? (3 marks)
(iii) Suggest **one** reason why the acceleration calculated in (ii) is not 10 m/s$^2$. (1 mark)

(Total: 16 marks)
(LEAG, Syll. A, Paper 3)

(a) See Unit 2.8
(b) (i) The flea accelerates for 0.001 s at the rate of 800 m/s$^2$. Use $v = u + at$ to calculate $v$ where $u = 0$.
(ii) The acceleration is found by calculating the gradient of the graph for the 67 ms. The acceleration is downwards.
(iii) Resistance of the air.

### Unit 3

2 Explain the fact that a body on the moon has the same mass as, but would register less on a spring balance than, the same body on earth. Explain why the spring balance shows this difference and why a lever arm balance would not.

Mass is a measure of the quantity and type of material in a body, and neither of these are affected by the body's position in the Universe. The spring balance stretches in proportion to the force of attraction of the earth or moon for the body (weight). These forces are different.
A lever arm balance uses the principle of moments to compare the weight of the body with

that of another body. Both bodies are attracted by either the earth or the moon, and no difference shows.

3 What is meant by the momentum of a body? State the law of conservation of linear momentum, including the conditions under which it applies.
   Describe an experiment which illustrates the validity of the law for simple collisions.
   Two bodies of masses 200 kg and 100 kg travel towards each other with velocities of 20 m/s and 25 m/s, respectively, and join to form one body on collision.
   (i) What is their common velocity after the collision?
   (ii) In what direction do they move after the collision?

The momentum of a body is the product of its mass and its velocity. It is a measure of how difficult it is to alter the motion of the body. The law of conservation of linear momentum states that, in a collision, the sum of the separate momenta of all the bodies, in a given direction, is the same afterwards as it was before. For this law to apply there must be no other forces acting on the bodies concerned.
See Unit 3.6.
Equation (3.8) in Unit 3.6 expresses the conservation law mathematically. The direction of travel of one of the colliding bodies is taken as positive and both bodies have the same velocity after collision, which can be calculated from the equation.

4 A front-wheel drive car is travelling at constant velocity. The forces acting on the car are shown in the diagram. Q is the force of the air on the moving car and P is the total upward force on both front wheels.

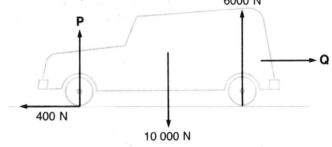

(a) Explain why (i) P = 4000 N, (ii) Q = 400 N. (2 marks)
(b) Calculate the mass of the car. (1 mark)
(c) The 400 N driving force to the left is suddenly doubled.
   (i) Calculate the resultant force driving the car forward.
   (ii) Calculate the acceleration of the car.
   (iii) Draw a sketch graph showing how the velocity of the car changes with time. (Start your graph just before the driving force is doubled.) (5 marks)
(d) (i) Passengers in a car are advised to wear a safety belt. Explain, in terms of Newton's laws, how a safety belt can reduce injuries.
   (ii) What other design feature in a car can offer protection in a crash? (4 marks)
(Total: 12 marks)
(WJEC, Paper 2)

(a) (i) 6000 N of the weight of the car are supported on the back wheels. The other 4000 N (10 000 N − 6000 N) must be supported on the front wheels.
   (ii) The car is travelling at constant velocity. Thus the driving force (400 N) must be equal but opposite to the force due to air resistance (Q).
(b) The mass of the car is found by using: $weight$ (10 000 N) = $mg$.
(c) (i) The extra 4000 N of force now accelerates the car.
   (ii) Use $F = ma$.
   (iii)

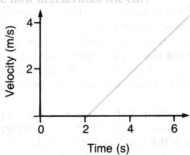

(d) (i) If passengers are not wearing a safety belt when the car stops suddenly in a crash, they will continue to move forward at their original speed and will collide with the front of the car.
   If they are wearing a safety belt, they will slow down more gently (over a longer time) as the belt 'gives' and the decelerating force on them will be less than if they had collided with the front of the car.
   (ii) A 'crumple zone' crumples on impact absorbing much of the decelerating force. This means that the force acting on the rest of the car and the passengers is less.

5 The kinetic energy of a moving car is transformed during the braking of the car. The distance over which this energy transformation occurs is called the **braking distance**. The **stopping distance** is greater than the braking distance because of the additional **thinking distance**. This is the distance the car travels at constant speed while the driver reacts to the need to brake.

(a) The table below is based on the Highway Code and gives the thinking, braking and stopping distance for a 1000-kg car.

| Speed km/h | Thinking distance/m | Braking distance/m | Stopping distance/m |
|---|---|---|---|
| 50 | 9 | 14 | 23 |
| 80 | 15 | 38 | 53 |
| 110 | 21 | 74 | 95 |
| 160 | 30 | 155 | 185 |

(i) Draw a graph of **braking** distance (y-axis) against speed (x-axis). (5 marks)
(ii) Is it sensible to extend (extrapolate) this graph to the origin? Give a reason for your answer. (2 marks)
(iii) If a graph of **thinking** distance against speed is drawn it is a straight line through the origin. Why are thinking distance and speed related in this way? (2 marks)

(b) When this 1000-kg car is travelling at 20 m/s (72 km/h)
(i) calculate the kinetic energy of the car, (2 marks)
(ii) read from the graph the braking distance for this speed, (1 mark)
(iii) calculate the average braking force during the braking from this speed. (2 marks)

(c) A typical car is of length 4.2 m. How many car lengths would it be suitable for a driver in such a car travelling at 100 km/h on a motorway to leave between himself and the car in front? (2 marks)

(Total: 16 marks)
(LEAG, Syll. A, Paper 3)

(a)(ii) Yes. It will take zero distance to stop a car which has zero speed.
(iii) The thinking (reaction) time is the same whatever the speed
(b)(i) Use $\tfrac12 mv^2$. (ii) 30.8 m
(iii) Use $force \times distance = kinetic\ energy$.
(c) The stopping distance is 95 m. This is 95/4.2 = 23 car lengths.

6 When the driver of a car sees an obstruction or hazard ahead she applies the brakes after a short (thinking) time. The table below shows how the distance travelled during this reaction time, the distance travelled after the brakes have been applied and the total stopping distance depend on the initial speed of the car.

| Initial speed m/s | Reaction (thinking) distance m | Braking distance m | Total stopping distance m |
|---|---|---|---|
| 15 | 9 | 15 | 24 |
| 25 | 15 | 42 | 57 |
| 35 | 21 | 82 | 103 |

(a) Use the information in the table to explain why
(i) the reaction time is 0.6 s, (1 mark)
(ii) the deceleration during braking is 7.5 m/s². (2 marks)
(b) Calculate the total stopping distance if the initial speed of the car is 30 m/s. (3 marks)
(c) Plot a graph showing how the total stopping distance depends on the initial speed. Use your graph to find the maximum speed from which the car could be stopped within 50 m. (3 marks)
(d) The car and driver of total mass 800 kg crash into a wall at a speed of 15 m/s. They rebound from the wall at 5 m/s, the time of contact with the wall being 0.5 s.
Calculate the average force exerted by the wall on the car. Explain carefully how a crumple zone (an easily compressed section at the front of the car) can protect the driver. (3 marks)

(Total: 12 marks)
(WJEC, Paper 2)

(a)(i) The reaction time is constant and is found by dividing the thinking distance by the speed of the car, i.e. 9/15 = 0.6 s.
(ii) Use $v^2 = u^2 + 2as$, i.e. $0 = 15^2 + 2a \times 15$, giving $a = -7.5$ m/s².
(b) Use $v^2 = u^2 + 2as$, where $u = 30$ m/s, $v = 0$ and $a = -7.5$ m/s².
(d) Use $Ft = mv - mu$, where $u = 15$ m/s and $v = -5$ m/s.
The crumple zone crumples in a collision thus increasing the time the car and driver take to stop. This reduces the force acting on them.

7 (a) Explain the meaning of the terms work, energy and power, stating the units in which each is measured.
(b) Describe in detail an experiment you could do to measure the average power you are able to develop over a few seconds.
(c) A crane whose engines can develop a power of 3 kW is used to raise a 500 kg load from the ground to a height of 12 m.
(i) Calculate the work done in raising the load through 12 m.
(ii) Calculate the minimum time for the crane to raise the load through this height.

(a) See Units 3.7, 3.8 and 3.10.

(b) Time pupils running up stairs. Average power $= \dfrac{mgh}{t}$ where $mg$ is the weight of the pupil, $h$ the height of the stairs and $t$ the time taken to run up the stairs.

(c) (i) Work done $= mgh = 500 \times 10 \times 12 = 60\,000$ J.

(ii) Minimum time $= \dfrac{\text{work done}}{\text{power}} = \dfrac{60\,000}{3000} = 20$ s. This assumes that the entire system is 100% efficient. Any loss of energy will result in the time taken being longer.

## Unit 4

8 Explain the expressions moment of a force about an axis, centre of mass.

Describe how you would find, experimentally, the position of the centre of mass of a flat sheet of wood of irregular outline.

A uniform plank AB of wood 2 m long, weighing 54 N, rests on a knife-edge 0.50 m from B. The end A is supported by a vertical string so that AB is horizontal.

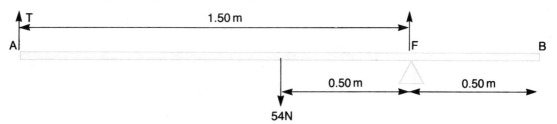

Find (a) the tension $T$ in the string, (b) the force $F$ acting on the knife-edge.

See the beginning of Unit 4 and Unit 4.2.
See Unit 4.3.
Take moments about the knife-edge, equating the moment due to the weight of the plank acting at a distance of 0.50 m from it to the moment of the string acting at a distance of 1.50 m from it.

The force $F$ acting on the knife-edge is the difference between the weight of the plank and the tension in the string. This is because the total vertical force on the plank must be zero as the plank does not move.

9 The three diagrams below illustrate a toy which always returns to the upright position when released. On diagram A, indicate by a letter $X$ a possible position of the centre of gravity of the toy.

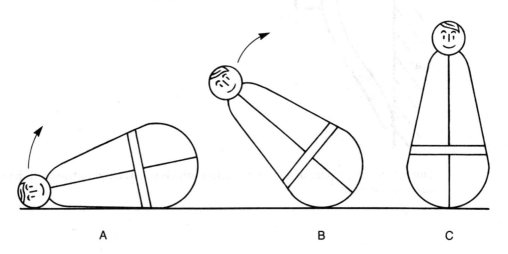

The centre of mass must be to the right of a vertical line through the point of contact of the toy with the ground in A and B. If this is so the toy's weight exerts a clockwise moment which rotates it into an upright position.

## Unit 6

10 The diagram illustrates the principle of the hydraulic brake used in many cars. The effective length of the brake pedal is 14 cm and the rod is attached 4 cm from the pivot.

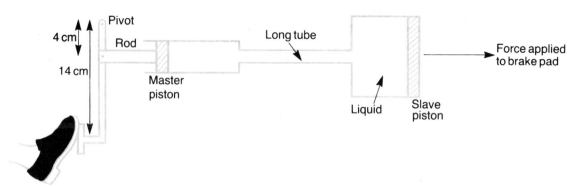

(i) What is the force applied to the master piston when an effort of 200 N is applied to the brake pedal?
(ii) If the area of the slave piston is three times that of the master piston, what force is applied to the brake pad?
(iii) What is the overall mechanical advantage of the system? (Assume that the system is 100% efficient.)

(i) See Unit 4.5. The velocity ratio is 3.5. Assuming 100% efficiency the force applied to the master piston is 700 N.
(ii) The liquid transmits the pressure on the master piston to the slave piston. Here it acts over three times the area, thus applying a force three times as great as at the master piston. Use the formula

$$\text{Pressure} = \frac{\text{force}}{\text{area}}$$

for solution.
(iii) The force applied increases from 200 N to the pedal to 2100 N at the pad. The mechanical advantage is $\frac{2100}{200} = 10.5$.

11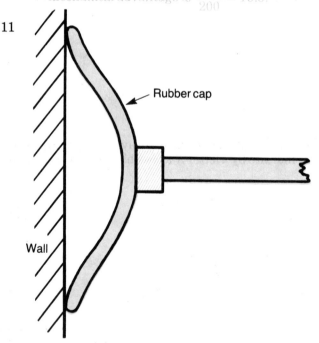

The figure shows part of a rubber sucker which is attached to the wall. Explain why it stays against the wall.

There is little air between the wall and the sucker as the air is squeezed out on application of the sucker to the wall. Inside the sucker the pressure is low whereas atmospheric pressure acts outside. This pressure difference holds the sucker to the wall.

## Unit 7

12 An experiment is performed to measure the increase in length of a spring as it is loaded with masses. The diagram shows the apparatus used and the graph shows the results obtained.

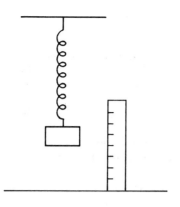
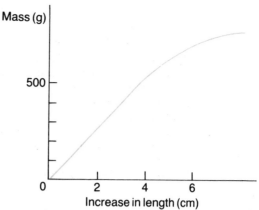

(a) Comment on the following statement. The apparatus could be used, together with the graph, to measure a downward force of about 5 N but might not be suitable for measuring greater forces.

(b) Given two springs of the same kind, they could be arranged either in series or in parallel, as shown in the diagrams.

(i) You load each of these arrangements with masses. Sketch a graph with lines, showing the result you would expect for each of these two arrangements.

(ii) Which arrangement would be more suitable for measuring a force of about 0.5 N? Give a reason.

(iii) Which arrangement would be more suitable for measuring a force of about 9 N? Give a reason.

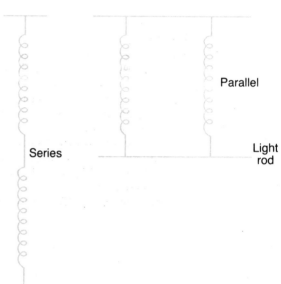

(a) The force due to gravity acting on a mass of 500 g is 5 N. The graph is straight up to this value showing that the spring is behaving elastically, and can be used to measure such a force. It would not be suitable for greater forces as these would permanently lengthen this spring.

(b) (i) The series arrangement is effectively a spring of twice the length. Thus twice the extension is obtained for corresponding forces compared to the single spring. The gradient is half of the original graph.

The parallel arrangement means that each spring experiences only half the total force applied. Thus half the extension is obtained for corresponding forces compared to the single spring. The gradient is twice that of the original graph.

(ii) The series arrangement gives a greater extension per unit force and is thus more suitable for small forces.

(iii) The parallel arrangement will give a straight line graph up to 10 N and is more suitable.

## Unit 8

13 Describe the differences between solids, liquids and gases in terms of (i) the arrangement of the molecules throughout the bulk of the material; (ii) the separation of the molecules; and (iii) the motion of the molecules.

See Unit 8.2.

14 State and explain, briefly, how the rate of evaporation of a pool of water is affected by (a) the area of surface exposed to the atmosphere and (b) the humidity of the atmosphere.

(a) If the pool has a large area of surface exposed to the atmosphere, then a higher proportion of its molecules are near the surface. Thus a higher proportion of the more energetic ones are able to escape from the surface. The rate of evaporation will be high.

(b) If the humidity of the atmosphere is high, it already contains a high concentration of

15 The botanist, Robert Brown, while examining pollen particles floating on water, noticed through his microscope that they were in a continuous state of rapid, random motion. When the stopper is removed from a bottle containing liquid ammonia, the ammonia, after a time, can be smelled in all parts of a room.
(a) Show how these and similar observations lead to a 'kinetic' theory of the structure of liquids and gases.
(b) Outline briefly the kinetic theory of gases and liquids.
(c) Show how the theory accounts for the pressure exerted by a gas on the walls of its container.
(d) Show how the theory accounts for the evaporation of a liquid.

(a) See Units 8.3 and 8.4. The smell must be due to some ammonia having moved across the room. This means liquid ammonia must consist of small particles (molecules) some of which have escaped. Also gaseous ammonia must consist of molecules which are moving. The fact that the smell takes some time to reach all parts of the room is due to the ammonia molecules continually colliding with air molecules on their way, which greatly reduces their average velocities.
(b) See Unit 8.1.   (c) See Unit 8.1.   (d) See Unit 8.1.

## Unit 9

16 The diagram illustrates a thermostat used to control an electric immersion heater.
(a) A and B are two metallic strips. State an important difference between them.
(b) Explain how the thermostat controls the temperature of the water heated by the electric immersion heater.
(c) What is the effect of rotating screw C?

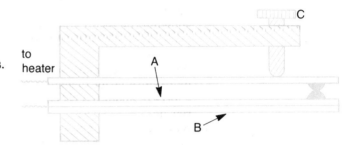

(a) The two strips expand by different amounts when the temperature increases, A more than B.
(b) The two strips are rivetted together and therefore when the temperature increases the combined strips bend downwards thus breaking the electrical contact at the right-hand end. This cuts off the current to the immersion heater. So when the water reaches the correct temperature, the electric current is cut off. When it cools a little the combined strips straighten and the current flows again.
(c) The screw C alters the temperature at which the thermostat operates. When it is tightened the water will have to be hotter before the combined strips bend sufficiently for the contact to be broken.

## Unit 10

17 (a) Name and state the law which defines the relationship between the volume and pressure of a fixed mass of gas at constant temperature.
(b) With the aid of labelled diagram, describe the structure and operation of the apparatus you would use to investigate this relationship.
(c) A certain mass of dry air has a volume of 1000 cm$^3$ when its pressure is 160 mm Hg below atmospheric pressure. Calculate its volume when its pressure is 40 mm Hg above atmospheric pressure and its temperature is unaltered.

(a) Boyle's law. See Unit 10.1.
(b) See Unit 10.1.
(c) Apply $PV$ = constant — Unit 10.1.

18 (a) (i) Describe how you would investigate the relationship between the volume and temperature of a fixed mass of dry air at constant pressure.
(ii) Draw a graph showing the results you would expect, labelling your axes.
(iii) Suppose you then used another pure gas (e.g. hydrogen or helium). What feature would the two graphs have in common?
(b) The following results were obtained in an experiment to study the change in the length of an air column with temperature.

| Temperature/°C | 0 | 40 | 80 |
|---|---|---|---|
| Length of air column/cm | 10.0 | 11.4 | 12.8 |

What would be the length of the air column at (i) 100°C, (ii) −100°C?

(a) (i) See Unit 10.2.
 (ii) The graph is a straight line which when extended to zero volume cuts the temperature axis at −273°C approximately.
 (iii) Both graphs when extended to zero volume cut the temperature axis at the same point (−273°C).
(b) Use $V/T$ is equal to a constant. The length of the air column is proportional to the temperature on the Kelvin scale ($T$). Use this relationship to calculate the new lengths of the air column.

## Unit 12

19 Explain
 (i) why a kettle of water with a steady supply of heat takes a much longer time to boil dry than it does to reach its boiling point;
 (ii) how evaporation differs from boiling; and
 (iii) how the molecular theory of matter accounts for the drop in temperature which results when rapid evaporation of a volatile liquid occurs.

(i) The latent heat of vaporization of water is greater than the heat required to raise the temperature of 1 kg of water by about 80°C (80 × specific heat capacity). It thus takes longer for the kettle to boil dry than to reach its boiling point.
(ii) See Unit 12.5.
(iii) See Unit 12.5. When evaporation is rapid the liquid has no time to absorb heat from the surroundings. The molecules left have a lower average kinetic energy, because the faster molecules have escaped.

20 (a) Explain the following in terms of molecules:
 (i) the Brownian motion of smoke particles; (2 marks)
 (ii) the cooling effect which occurs when a liquid evaporates; (2 marks)
 (iii) the transfer of heat by convection from a hot radiator to other parts of a room. (2 marks)
(b) An electric shower unit is rated at 7 kW. Cold water from the water main enters at 12°C and emerges as hot water at the rate of 4.0 kg per minute.
 (i) If the specific heat capacity of water is 4200 J/kgK, calculate the temperature of the hot water. (2 marks)
 (ii) By making a sensible guess at the quantities involved, compare the cost of taking a shower and having a bath. (2 marks)
 (iii) Describe **two** features of such an electric shower necessary for the safety of a user. (2 marks)
(Total: 12 marks)
(WJEC, Paper 2)

(a) (i) See Unit 8.3.
 (ii) See Unit 12.5 and Question 19 part (iii) above.
 (iii) When the air alongside the radiator becomes warmer its molecules move around more energetically. The density of the warmed air falls, so the warmed air rises. Colder air, which is more dense, falls to take its place, setting up convection currents.
(b) (i) Energy is being supplied at the rate of 7000 J/s.
 Rate of supply of energy $= \frac{mcT}{t} = \frac{4 \times 4200 \times T}{60}$, so $T = \frac{7000}{280} = 25°C$.
 $T$ is the temperature increase. So the final temperature is 37°C.
 (ii) A shower takes about 5 minutes i.e. about 20 kg of warm water. A bath may take 300 kg of water at the same temperature. A bath will therefore cost 15 times as much as a shower.
 (iii) A 'cut-out' is required so that if the water temperature gets too hot for any reason, the water is cut off.
 The water supply must be extremely well insulated from the power supply to avoid electrocution.

## Unit 13

21 (a) (i) On a cold night in winter condensation occurs on the inside of the windows of a car. Explain carefully why this happens. Would you expect the effect to be greater if there were four people in the car instead of one? Give your reasons.
 (ii) House windows are often made of two sheets of glass with an air gap in between (double glazing). State, with reasons, what effect you would expect this to have on the condensation on the surface of the glass facing into the room.
 (iii) If it were possible to remove the air from between the two sheets of glass what difference would you expect? Give reasons.
 (iv) State and explain two other beneficial effects which double glazing has over a single sheet of glass.

(b) Why is frost more likely to occur on a clear night than on a cloudy night? Explain why, under these conditions, the frost is usually more severe in lower lying areas.

(a) (i) The air in contact with the windows is at a lower temperature than the rest of the air in the car. The cooler air has a lower saturated vapour pressure and is unable to contain all the water vapour in it; some condenses out. With four people in the car there will be more water vapour in the air than with one, so the effect will be greater.
(ii) The air gap reduces conduction through the window, as air is a poor conductor. The inner plane of glass remains almost at room temperature, so less condensation occurs here.
(iii) Even less conduction would occur due to the complete lack of a medium. Any convection currents in the gap would also be eliminated. Even less condensation would take place.
(iv) It reduces heat losses and cuts down the noise entering the house.
(b) On a clear night the earth loses a great deal of heat by radiation, whereas on a cloudy night this heat is reflected from the clouds. Cold air is more dense than warm air; thus the temperature in low lying areas is lower.

22 Explain the following in terms of conduction, convection and radiation of heat.
(a) Houses in hot sunny countries traditionally have thick, white painted walls, with small windows.
(b) A vacuum flask keeps tea hot.
(c) Glider pilots are able to find upward moving currents of air over certain features such as towns.
(d) The blackboard in a class room stays at the same temperature as the rest of the room, even though it absorbs more radiant heat-energy than the light coloured walls.
(e) On a cool day iron railings feel colder to your touch than a wooden bench.

(a) Thick walls reduce the heat which enters the house as a result of conduction through the walls. If the walls are white a small fraction of the energy falling on them is absorbed, most is reflected. Small windows mean that little heat enters by radiation.
(b) The vacuum between the walls reduces the energy leaving by conduction or convection, as both these require a material through which to travel. Silvering of the walls facing the vacuum reduces the radiation which crosses this gap.
(c) Warm air has a lower density than cold air and hence rises. The gliders rise on the warm convection currents so caused.
(d) The blackboard absorbs more radiation from the surroundings than the white walls, but also radiates more. The temperature it acquires depends on a balance between its absorption and radiation and this is the same temperature at which the absorption and radiation of the white walls balance.
(e) Heat is conducted from the hands to the interior of the rails rapidly and this gives the sensation of cold. For wood the conduction is considerably less and hence the sensation of cold is less, making the wood feel warmer.

## Unit 14

23 (a) For each of the following, name a different electromagnetic wave and write a sentence describing how it is used.
(i) An electromagnetic wave being used to transfer energy. (2 marks)
(ii) An electromagnetic wave being used to send speech information. (2 marks)
(iii) An electromagnetic wave being used to measure distance. (2 marks)
(b) All waves can be diffracted. Describe briefly an experiment to show diffraction for one type of wave. Draw a diagram of the apparatus needed. (2 marks)
(c) Water waves on a ripple tank are often used to demonstrate the properties of waves. A single dipper vibrates up and down at a constant frequency. It is placed in the middle of a ripple tank and the pattern of waves produced is shown in **A**.

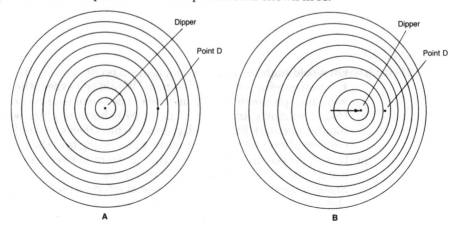

A wave detector at point D receives the waves at a frequency of 20 Hz.
(i) In diagram **A**, how long has it taken a wave to travel from the dipper to point D?
(3 marks)

A dipper (still vibrating) is moved across the ripple tank from left to right at a steady speed so that each wave starts from a different place in the tank. The speed of the waves across the water is not affected by his movement. The pattern produced is drawn in **B**.

(ii) Measure diagram **B** and diagram **A** to find out the effect this movement of the dipper has on the wavelength of the waves received at **D**. What is the new frequency? (Both diagrams are drawn to the same scale.) (4 marks)

(iii) In diagram **B**, what is the frequency received at D after the dipper has passed D and is moving away? (2 marks)

(Total: 17 marks)

(MEG, Syll. A, Paper 3)

(a) (i) heat or light  (ii) radio  (iii) radar
(b) See Unit 14.6.
(c) (i) 20 waves are made each second. The wave at D is 5 waves from the dipper. It has thus taken 0.25 s to get there.
(ii) There are now 9 waves in 1.8 cm rather than 9 waves in 2.6 cm.
(ii) The new frequency passing D is thus $\frac{20 \times 2.6}{1.8} = 29$ Hz.
(iii) The new frequency is $\frac{20 \times 2.6}{3.45} = 15$ Hz.

## Unit 15

24 Describe the construction of a pinhole camera. Upon what property of light does it rely for its action? Discuss the effect on the size, sharpness and brightness of the image of
(a) doubling the object distance;
(b) doubling the diameter of the hole.

See Unit 15.1. The pinhole camera relies on the fact that light travels in straight lines.
(a) Halves the size and makes the image sharper but dimmer.
(b) The size remains about the same, but the image is much more blurred. It is also brighter.

25 Explain with the help of a diagram how the eye sees the image of a bright point formed by a plane mirror.

A man sits in a chair looking into a plane mirror which is 2 m away from him. How far away from him is the image of his face.

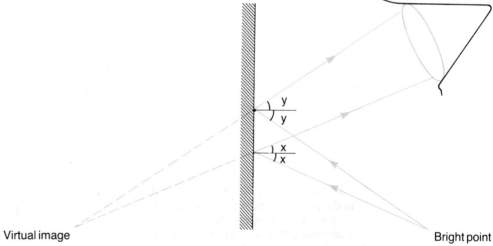

Virtual image                                    Bright point

The image formed by a plane mirror is always as far behind the surface of the mirror as the object is in front. The image of the man's face is thus 4 m away from him.

## Unit 16

26 The diagram shows water waves passing from deep to shallow water in a ripple tank.
(a) Copy the diagram and add the waves which have already passed into shallow water.
(b) Describe what happens to the waves as they go into shallow water. Account for their behaviour.

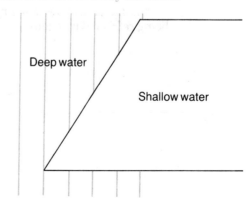

The diagram below shows a parallel beam of white light passing through a prism.

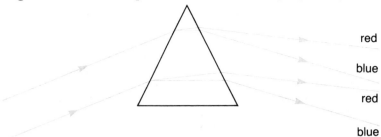

(c) Describe the following effects in terms of waves.
(i) The light changes direction on entering and leaving the prism.
(ii) The white light splits up into different colours in passing through the prism.
(d) Explain how you could use a second prism to recombine the colours.

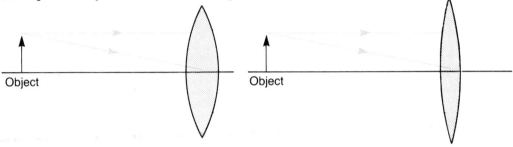

(e) Copy the diagrams above and show the rays after they have passed through each lens. In both cases the lens forms a real image.

(a) See Fig. 14.6.
(b) See Unit 14.4.
(c) (i) Light travels more slowly in glass than in air. The light changes its direction of travel in a similar way to the water waves in (a).
(ii) White light is a mixture of many wavelengths (colours). The shorter wavelengths slow down more than the longer wavelengths on entering glass. Thus the shorter wavelengths (e.g. blue) change direction more than the longer wavelengths.
(d)

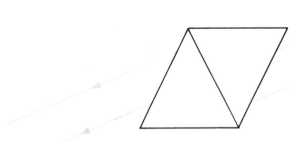

(e) See Fig. 16.9(c).

27 The diagram shows water waves moving from deep to shallow water and returning to deep water in a ripple tank.
(a) Copy the diagram adding the waves in the shallow water and in the deep water beyond the shallow region.

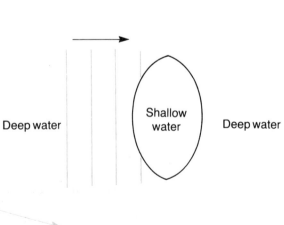

This diagram shows a beam of red light being refracted by a lens.

(b) How does the behaviour of water waves help to explain this refraction?
(c) Explain why white light splits up into colours when passing through a lens.

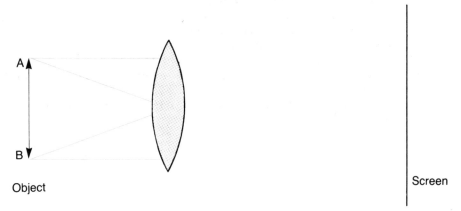

Object                      Screen

An object AB is placed in front of a lens as shown in the diagram. Two rays have been drawn from A to the lens and two from B to the lens.

(d) Copy the diagram and continue the four rays beyond the lens to show the formation of the image of AB on the screen.
(e) Comment on the nature of the image.
(f) State what happens to the position and size of the image if the object is moved away from the lens.
(g) Draw a ray diagram to show how a converging lens is used as a magnifying glass. Label the position of the image.

(a)

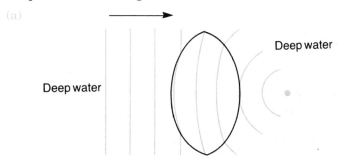

(b) Water waves slow down on passing into shallow water; light does the same on passing into glass. In both cases that part of the wave which reaches the shallow water or glass first, slows first. The waves change their direction of motion and converge. The centre of the wave leaves the shallow water or glass last and speeds up last. Thus more convergence occurs.
(c) See Question 26 part (c)(ii).
(d)

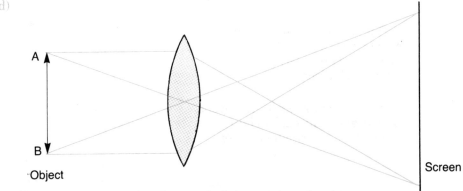

(e) The image is inverted and magnified (larger than the object).
(f) The image is closer to the lens and smaller.
(g) See Fig. 16.9(a).

## Unit 17

28 The behaviour of light has been explained by using a particle model and by using a wave model.
    (a) How is refraction explained using a particle model? (2 marks)
    (b) How is refraction explained using a wave model? (3 marks)
    (c) What test was used to decide which was the better model to use to explain refraction, and what was the result of the test? (2 marks)

(d) A book says 'If red light is passed through two narrow slits close together, fringes appear on a screen placed beyond the slits. This is called interference.'
Use the wave model of light to explain
(i) why there are bright places in the fringes;
(ii) why there are dark places in the fringes;
(iii) why there can be many fringes. (6 marks)
(e) Why cannot the particle model explain interference? (2 marks)
(f) The colours seen in a soap film, or in a film of oil on water, in daylight, are due to interference.
(i) Explain how interference can take place in the case of thin films. (3 marks)
(ii) Why are colours seen? (2 marks)
(Total: 20 marks)
(MEG, Nuffield, Paper 3)

(a) When a ray of light enters glass from air it bends towards the normal (inwards). If light is thought of as a beam of particles, then the particles would have to increase in speed for this to happen.
(b) See Question 27 part (b).
(c) When the speed of light in glass is measured it is found to be less than in air. This evidence supports the wave model.
(d) See Unit 17.2.
(e) It is not possible to conceive of two particles cancelling out to give a dark region.
(f) (i) When light falls on a thin soap film some light is reflected back from the front surface of the film and some from the back surface. These two reflected beams interfere in a similar way to light from two slits.
(ii) Daylight consists of a mixture of all the colours of the spectrum. When the film is viewed at an angle the path difference for light reflected from different parts of the film differs. The various colours therefore add up at different parts of the film as seen by the eye.

29 Straight water waves in a ripple tank move towards a gap in a straight barrier as shown in the diagram.
(a) Copy the diagram and draw in four waves which have passed through the gap in the barrier.
(b) Sketch how the four waves beyond the gap would appear if the gap were very much wider.

This diagram shows straight water waves in a ripple tank moving towards two gaps in a straight barrier.

(c) Sketch the wave pattern beyond the two gaps, and explain it in terms of the behaviour of waves.

A lamp is set up so that its bright, straight filament is vertical. A red filter is placed in front of the lamp, and a cylindrical lens is placed between the lamp and the screen so that a sharp image of the filament is seen on the screen.
(d) A blackened sheet of glass with two very narrow, close, and parallel slits ruled on it is put near the lens as shown in the diagram (page 207). What will be seen on the screen?
(e) What would be the effect of replacing the red filter in (d) by a blue one?
(f) What would be the effect in (d) of covering one of the slits?

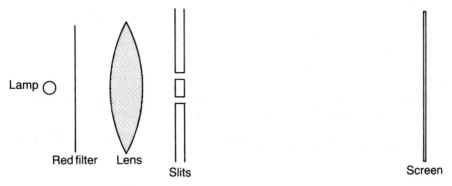

(a) See Fig. 14.9(b)   (b) See Fig. 14.9(a)   (c) See Fig. 14.7.
The waves spread out on passing through each of the slits. Beyond the slits the two sets of waves interfere. Where crest meets crest and trough meets trough the waves add up to give large crests and troughs. Where trough meets crest the two waves cancel completely to give calm water.
(d) Bright (red) and dark vertical parallel interference fringes are seen.
(e) The bright fringes would be blue and the fringes would be closer together.
(f) There would be one patch of light on the screen — no interference fringes.

## Unit 18

30

| radio | micro waves | (p) | visible | ultra violet | (q) | gamma |
|---|---|---|---|---|---|---|

(a) The sketch above shows part of the electromagnetic spectrum.
Name
(i) the missing types of radiation (p) and (q),
(ii) the radiation with the smallest wavelength,
(iii) the radiation with the highest frequency,
(iv) a method of detecting ultra violet,
(v) the origin of gamma radiation.                                                  (3 marks)
(b) With the help of a labelled diagram in **each** case, explain what is meant by
(i) refraction of waves,   (ii) diffraction of waves.                                (4 marks)
(c)

When the wire was plucked and the tuning fork struck, the tuning fork had the higher pitch.
(i) State **two** ways by which the wire could be made to produce the same note as the tuning fork.                                                                                     (2 marks)
(ii) State **one** way in which the vibration of the string and the sound waves in the air are the same, and one way in which they are different.
(d) A young scientist said 'I think thunder and lightning should really be called lightning and thunder'. Why does this seem a sensible suggestion?                             (1 mark)
(Total: 9 marks)
(WJEC, Paper 2)

(a) (i) (p) infrared   (q) X-rays   (ii) gamma   (iii) gamma   (iv) fluorescence
(b) (i) See Fig. 14.6.   (ii) See Fig. 14.9.
(c) (i) Shorten the wire, e.g. by putting a finger on it at a suitable point.
    Increase the tension in the string by attaching a larger weight.
  (ii) They have the same frequency. However, the vibration of the string is transverse whereas the sound waves in the air are longitudinal.
(d) The lightning is normally seen before the thunder is heard because light travels faster than sound.

31 In a cathedral, a man sitting 165 m from the organ and 5 m from the nearest loudspeaker notices a time delay of 0.5 s between the start of a chord transmitted by the amplification system and that transmitted through the air in the cathedral. Calculate the velocity of sound in the air in the cathedral and state what assumption must be made to enable this to be done.

## Unit 21

32 (a) Draw a diagram of the circuit you would use to investigate how the current through a conductor depends on the potential difference applied to it. (2 marks)

(b) The graphs below show the results of two such investigations, one using a car headlight bulb and the other using 60 cm of nichrome wire.

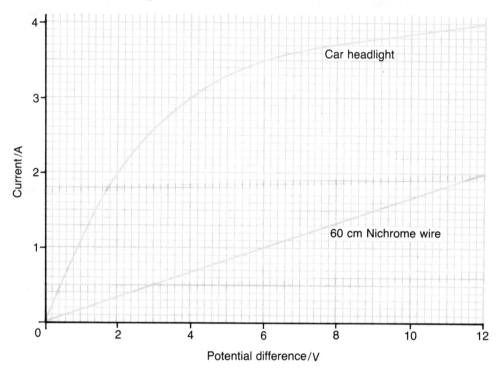

Using the information on the graph,
(i) calculate the resistance of the lamp and wire when the potential difference is 12 V.
(ii) calculate the resistance of the lamp and wire when the potential difference is 2 V.
(iii) Comment on the results you have obtained in (i) and (ii).
(iv) Calculate the resistance of 1 m of nichrome wire.
(v) What length of nichrome wire would pass a current of 4 A if a potential difference of 12 V were applied to it? (6 marks)

(c) The lamp and 60 cm wire were then connected in the following circuits.

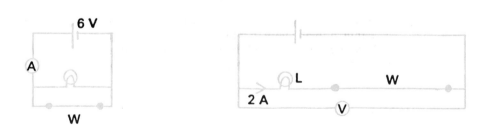

Use the graph to find the readings on the ammeter and voltmeter. (4 marks)
(Total: 12 marks)
(WJEC, Paper 2)

(a) See Fig. 21.3.
(b) (i) For car headlight, $R = V/I = 12/4 = 3\,\Omega$; for nichrome wire, $R = V/I = 12/2 = 6\,\Omega$.
(ii) For car headlight, $R = 1\,\Omega$; for nichrome wire, $R = 6\,\Omega$.
(iii) The resistance of the car headlight increases as more current passes through it because it gets hot. The resistance of the nichrome wire is constant.

(iv) 60 cm has a resistance of 6 Ω, so 1 m will have a resistance of 10 Ω.
(v) Using $R = V/I$, the resistance of this wire would be 3 Ω. 30 cm of wire will have a resistance of 3 Ω.
(c) Left-hand circuit: Given the p.d. (6 V), read the current values from the graph. The ammeter will read 4.5 A (1 A plus 3.5 A).
Right-hand circuit: Given the current (2 A), read the p.d.s from the graph. The voltmeter will read 14 V (2 V and 12 V).

33 Two heating coils dissipate heat at the rate of 40 W and 60 W, respectively, when connected in parallel to a 12 V d.c. supply of negligible internal resistance. Calculate the resistances of the coils.
   Assuming that these resistances remain constant, what would be their rates of dissipation of heat when connected together in series with the same supply as before?

See Unit 21.8. Apply *current = power/potential difference* in each case and then apply $V/I = R$.
   The resistances add up when placed in series. Calculate the current passing through them using $I = V/R$. Then use power $= I^2R$ for each one.

34 In the following circuits, similar cells of negligible resistance, ammeters and resistors are used.

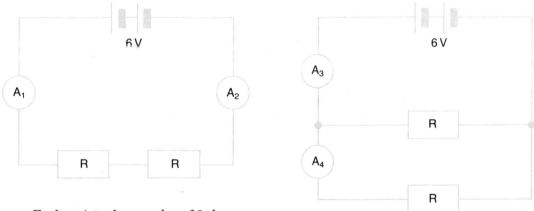

Each resistor has a value of 6 ohms.
(a) Calculate the readings of the ammeters numbered 1–4.
(b) Calculate the total power transformed in each circuit.

An electric iron is marked 240 V, 750 W.
(c) Explain the meaning of these markings.
(d) Calculate the current passing through the iron.
(e) Calculate the resistance of the heating element in the iron.

The heating element of this iron is made from two resistors, each with a working value of 153.6 ohms.
(f) Are the resistors connected in series or parallel?
(g) State one advantage of the arrangement that you have given in answer to (f).

When the iron is first switched on, the element warms up, but eventually it is kept at a steady temperature. This is achieved by a thermostat placed in the circuit inside the iron, as shown in the diagram.
(h) Explain how the thermostat works.
(i) How is heat from the element prevented from making the handle of the iron too hot to hold?
(j) How is the metal casing of the iron made safe so that there is no danger to the user should the live wire accidentally touch it?

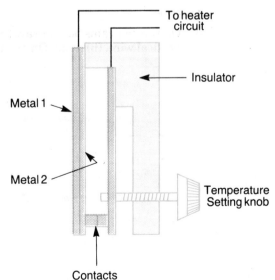

(a) The current in the series circuit is given by $\frac{V}{R+R}$. In the parallel circuit the current through each resistor is given by $\frac{V}{R}$.
(b) Use $I^2R$ to calculate the power in each circuit.
(c) See Units 21.1 and 21.7.
(d) Use Power $= VI$ to calculate $I$.
(e) Use $V = IR$ to calculate $R$ using the value of $I$ obtained in part (d).
(f) The value of the resistance calculated in (e) is half of 153.6 ohms. Two 153.6 ohm resistors in parallel will give this value.
(g) If one of the resistors in the element stops working the iron will continue to work on half power.
(h) See Unit 9.1.
(i) There is an insulator between the element and the handle. The handle is made from a poor conductor of heat.
(j) The metal casing is connected to the earth lead. See Unit 21.12.

35 (a) (i) Why would it be wrong to fit an electric fire in a bathroom on the wall directly above the bath?
   (ii) Where should such a heater be fitted and what type of switch should be used to operate it?
(b) The flex to the 13 A 3-pin plug shown in the diagram has been incorrectly fitted. List four mistakes that have been made.

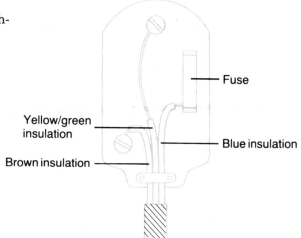

(a) (i) In this position it would be possible for someone standing in the bath water to grasp the live part of the fire, for example the element. They are also likely to have wet hands and thus to make good contact with the element and with earth through the water.
   (ii) It should be placed on a wall out of reach from the bath and should be operated using a switch with a hanging cord.
(b) (i) The live lead (brown) should be connected to the fuse, whereas here the neutral lead (blue) has been connected to it.
   (ii) In any case the blue lead has been connected to the wrong end of the fuse. As connected here the lead goes directly to the pin, and the fuse is not in the circuit.
   (iii) Too much insulation has been removed from the earth lead. There is a danger of it 'shorting' to the brown lead.
   (iv) The insulation round the entire cable has been stripped back too far. It should go under the cable clamp. As shown here any pull on the cable is taken by the three leads.

## Unit 22

36 On the diagrams below draw the magnetic field patterns for current flowing in (a) a straight vertical wire, (b) a coil. On each diagram use arrows to indicate the direction of the field.

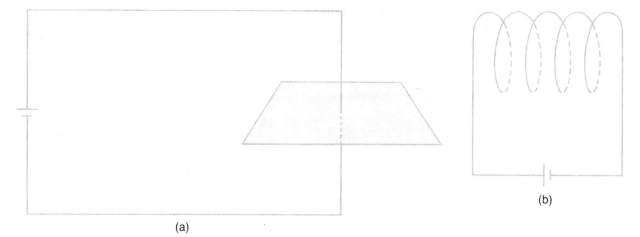

(a) See the beginning of Unit 22, also Figs 22.1 and 22.3.
(b) See the beginning of Unit 22.

37 (a) The diagram shows a rectangular current-carrying coil ABCD pivoted about a horizontal axle and placed between the poles of a magnet which has flat pole pieces.

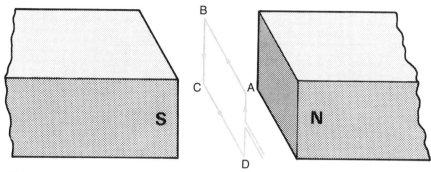

The current in the coil passes in the direction of the arrows.
 (i) Copy the simpler diagrams below and add to them labelled arrows to show the direction of the force, if any, on each of the four sides of the coil due to the interaction of magnetic fields. ($F_{AB}$ is a suitable label for the force acting on the side AB of the coil.) (4 marks)

 (ii) State whether or not with the coil in this position the coil will rotate about the axle. Explain why this is. (2 marks)
(b) When such a coil is used in a simple electric motor, a split-ring commutator may be used to change the current direction when necessary. Draw a simple diagram of such a commutator connected to a coil. In what position is the coil when the current direction is reversed? (3 marks)
(c) Such a coil may also be used in a moving coil meter. Here it is usual for the magnet to have curved pole pieces and a fixed iron core as shown in the diagram below.

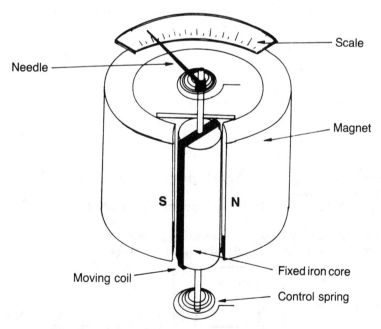

 (i) Draw a diagram to show the magnetic flux pattern between the pole pieces and the fixed iron core. (3 marks)
 (ii) When there is a current in the coil, it rotates and the needle moves across the scale. This coil rotation causes the control springs to tighten.
  1 Why does the needle stop moving although current is passing? (1 mark)
  2 Why does a larger current give a larger scale reading? (1 mark)

3 Why when the current is switched off does the needle return to the zero of the scale?
(2 marks)
(Total: 16 marks)
(LEAG, Syll. A, Paper 3)

(a) (i) See Unit 22.4.
  (ii) No. The forces are tending to pull the coil apart not rotate it.
(b) See Fig. 22.9. Horizontal.
(c) (i)

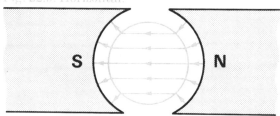

(ii) 1 When the coil starts to rotate it turns the springs, which then push back. Eventually the force due to the springs is as strong as that due to the current and the coil turns no further.
2 A larger current gives a larger force. The springs must therefore rotate through a larger angle before they push back with a force equal to that due to the current.
3 There is no longer a force on the coil and hence on the springs.

## Unit 23

38 The circuit shows a step-down transformer used to light a lamp of resistance 4.0 Ω under operating conditions.

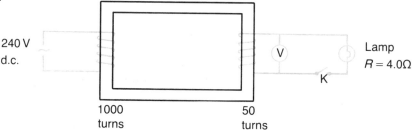

Calculate
(a) the reading of the voltmeter, V, with K open;
(b) the current in the secondary winding with K closed (the effective resistance of the secondary winding is 2.0 Ω);
(c) the power dissipated in the lamp;
(d) the power taken from the supply if the primary current is 150 mA; and
(e) the efficiency of the transformer.

(a) $\dfrac{\text{Secondary voltage}}{\text{Primary voltage}} = \dfrac{\text{number of secondary turns}}{\text{number of primary turns}}$. This is a step down transformer with a turns ratio of 20:1. The voltage is stepped down by the same factor.
(b) Apply $I = V/R$ where $R = 6.0$ ohms, using answer from (a).
(c) Power $= I^2R$, where $R = 4.0$ ohms.
(d) The input power = primary current × primary voltage.
(e) Efficiency $= \dfrac{\text{power output (secondary)}}{\text{power input (primary)}} = \dfrac{(c)}{(d)}$

39 (a) (i) Describe briefly, the energy changes involved in the generation of electrical energy at a power station. (2 marks)
  (ii) State and explain **two** advantages of a hydroelectric power station compared with other types. (2 marks)
(b) What are the advantages of transmitting power at
  (i) very high voltage?   (ii) alternating voltage? (3 marks)
(c)

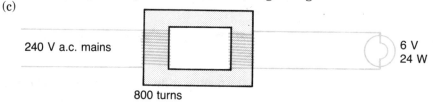

A 6 V, 24 W lamp shines at full brightness when it is connected to the output of a mains transformer. Assuming the transformer is 100 per cent efficient, calculate
(i) the number of turns in the secondary coil if the lamp is to work at its normal brightness,
(ii) the current which flows in the mains cable. (2 marks)

(d) Explain whether, and how, the number of secondary turns of the transformer shown in (c) should be altered if
(i) two 6 V lamps in **series** are to work at normal brightness,
(ii) two 6 V lamps in **parallel** are to work at normal brightness. (3 marks)
(Total: 12 marks)
(WJEC, Paper 2)

(a) (i) Chemical energy in the fuel (coal or oil) or nuclear energy in the reactor is converted to heat. This heat is then used to drive a turbine, i.e. it is converted into kinetic energy. The generator then converts the kinetic energy into electrical energy.
(ii) The source of energy, i.e. falling water, is renewable.
Hydroelectric power stations, unlike coal- and oil-fired ones, cause no pollution of the atmosphere.
Hydroelectric power stations can be used to store energy (pumped storage) at times of low demand.
(b) (i) If a fixed amount of power is transmitted at a very high voltage as opposed to a lower one, the current passing through the transmission lines is much lower. This greatly reduces the energy lost which depends on the current squared.
(ii) Transformers can be used to step up and step down the voltage thus reducing the energy losses in the transmission lines.
(c) (i) Using $\frac{N(\text{secondary})}{N(\text{primary})} = \frac{V(\text{secondary})}{V(\text{primary})}$, $N(\text{secondary}) = 20$.
(ii) I(secondary) = 24/6 = 4 A. At 100 per cent efficiency the power in the secondary coil will be the same as in the primary. As V is 40 times more in the primary than in the secondary, the current will be 40 times less, i.e. 0.1 A.
(d) (i) 12 V are required, therefore 40 turns are needed on the secondary coil.
(ii) 6 V are required; no change.

## Unit 24

40 How would you show that the particles emitted from the heated cathode of a thermionic diode are negatively charged?

A power supply is connected to the heater (normally 6.3 V d.c. or a.c.). A suitable potential difference is connected between the cathode and anode such that the anode is positive (Fig. 24.1a). A current flows. However, if, by reversing the connections of this power supply, the anode is made negative with respect to the cathode, no current flows.
If the heater is off no current flows in either case. Thus, when the heater is on, negatively charged particles must be leaving the cathode; these are attracted by the anode when it is positive, but repelled when it is negative.

41 A manufacturer makes an electron 'deflection tube' which consists of an evacuated glass bulb with some electrodes in it. He states that the tube works best if a 6 V supply is connected between A and B and a 3 kV supply between A and C. The electrode at C has a small hole in it, and just beyond the hole there are two parallel metal plates with a connection to each of them.
(a) What is the purpose of having a potential difference between A and B?
(b) Which terminal of the 3 kV supply should be connected to C?
(c) Explain the purpose of the high voltage.
(d) Why is the tube 'evacuated'?

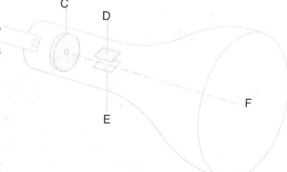

When the manufacturer's instructions are followed, a spot of green light appears at F on a special coating on the inside of this deflection tube.
(e) When a p.d. of 100 V is applied between D and E (D positive) the spot moves away from F. Draw a diagram to show the direction in which the spot moves.
(f) Explain why the spot moves in (e).
(g) How could you make the spot move further from F?
(h) If the spot is not very bright, what two separate changes might you make to increase the brightness?
(i) An alternating voltage of frequency 50 Hz is applied across the plates D and E. Describe what you would now see on the screen.

See Unit 24.3.
(a) The potential difference between A and B causes the filament to heat so that electrons are released.

42 (a) Name and draw the symbol for a logic gate in which
  (i) the output is high (1) when the input is low (0),
  (ii) the output is high (1) when both inputs are low (0). (3 marks)
  (b) A cooling fan in a busy office is switched on automatically when it gets **too hot** during the **daylight hours**.
  A possible arrangement might be

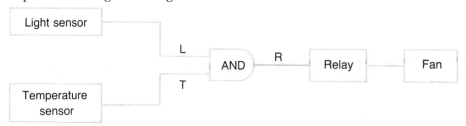

  (i) Name a suitable component used in
    (A) the light sensor, (B) the temperature sensor.
  (ii) Construct the truth table for the AND gate.
  (iii) What is the purpose of the relay? (4 marks)
  (Total: 7 marks)
  (WJEC, Paper 2)

(a) (i) NOT  (ii) NAND  See Fig. 24.12 for symbols.
(b) (i) A a light dependent resistor  B a thermistor
   (ii)

| L | T | R |
|---|---|---|
| 0 | 0 | 0 |
| 1 | 0 | 0 |
| 0 | 1 | 0 |
| 1 | 1 | 1 |

   (iii) The relay completes the circuit operating the fan and thus switches it on.

# Unit 25

43 (a) $^{23}_{11}$Na and $^{24}_{11}$Na are isotopes of sodium. $^{24}_{11}$Na is radioactive with a half-life of 15 hours.
  (i) Describe the structures of the atoms of the two isotopes and explain whether or not they can be separated by chemical means. (3 marks)
  (ii) Describe an experiment you would carry out to determine whether $^{24}_{11}$Na emits alpha, beta or gamma radiation. (4 marks)
  (iii) If $^{24}_{11}$Na does emit beta radiation, what effect would this have on the nucleus? (1 mark)
  (iv) Owing to a leakage of the radioactive isotope in a laboratory, the radiation count-rate rises to eight times the safety level. After how long will it be safe to enter the laboratory again? (2 marks)
  (b) Briefly describe **one** industrial **or** medical use of radioactivity. (2 marks)
  (Total: 12 marks)
  (WJEC, Paper 2)

(a) (i) Both isotopes of sodium contain the same number of protons in their nuclei and an equal number of peripheral electrons. They differ only in the number of neutrons in their nucleii. They cannot be separated chemically.
  (ii) See Unit 25.4.
  (iii) A neutron decays to a proton and an electron. The electron (beta ray) is ejected but the proton remains in the nucleus. The charge of the nucleus is thus increased by one.
  (iv) After three half-lives (i.e. 45 hours) have passed the count rate will have dropped by a factor of 8 and will be below the safety level again.

44 (a) Natural radioactive substances emit three kinds of radiation. Name the three types and state what each of them is. (3 marks)
(b) How would you show the different penetrative powers of the radiations? (6 marks)
(c) Use is made of radiation in hospitals. Describe briefly how it is used and for what purpose. (4 marks)
(d) Special precautions are necessary when radioactive substances are used or stored in a hospital. State **two** precautions and the reasons for them. (4 marks)
(e) When the Chernobyl disaster took place, certain radioactive isotopes fell on some parts of the United Kingdom. The activity increased above the background count.
(i) What is meant by an 'isotope'?
(ii) What is a 'radioactive' isotope?
(iii) What is meant by 'background count'?
(iv) It was said that, after some weeks, what had fallen would no longer be a hazard. How is it possible that it could no longer be dangerous? (5 marks)
(f) What is meant by **uranium fission**? How does this process produce heat in a nuclear reactor? (3 marks)
(Total: 25 marks)
(MEG, Nuffield, Paper 3)

(a) Alpha = helium nuclei; beta = electrons; Gamma rays = high frequency electromagnetic waves.
(b) See Unit 25.4.
(c) See answer to Question 43(b).
(d) See Unit 25.6.
(e) (i) Isotopes are atoms of the same element (same number of protons) with different numbers of neutrons.
(ii) One which gives out alpha or beta particles or gamma rays.
(iii) This is the level of radioactivity which is always present, due to such things as cosmic rays and radiation from the rocks in the earth's crust.
(iv) All radioactive substances decay and after a time the count rate falls to a relatively safe level.
(f) See Unit 25.7.

## Numerical answers to longer questions

1 (b) (ii) 11.9 m/s$^2$
3 (i) 5 m/s  (ii) in the same direction as the 200 kg mass was moving
4 (b) 1000 kg
  (c) (i) 400 N  (ii) 0.4 m/s$^2$
5 (b) (i) 200 000 J  (ii) about 30 cm
    (iii) about 6700 N
  (c) 23
6 (b) 78 m  (c) 23 m/s  (d) 32 000 N
7 (c) (i) 60 000 J  (ii) 20 s
8 (a) 18 N  (b) 36 N
10 (b) (i) 700 N  (ii) 6.5 cm
17 (c) 750 cm$^3$
18 (b) (i) 13.5 cm  (ii) 6.5 cm
20 (b) (i) 37°C

23 (c) (i) 0.25 s  (ii) 29 Hz  (iii) 15 Hz
25 4 m
31 320 m/s
32 (b) (i) headlight 3 Ω, wire 6 Ω
       (ii) headlight 1 Ω, wire 6 Ω
       (iv) 10 Ω  (v) 3 Ω
   (c) left 45 A, right 14 V
33 3.6 Ω and 2.4 Ω, 24 W
34 (a) $A_1$ = 0.5 A  $A_2$ = 0.5 A  $A_3$ = 2 A  $A_4$ = 1 A
   (b) 3 W, 12 W  (d) 3.125 A  (e) 76.8 Ω
38 (a) 12 V  (b) 2 A  (c) 16 W  (d) 36 W
   (e) 44.4%
39 (c) (i) 20 turns  (ii) 0.1 A
   (d) (i) 40 turns  (ii) 20 turns
43 (a) (iv) 45 hours

# INTERNAL ASSESSMENT OF PRACTICAL WORK

This carries at least 20% of the final marks for the examination (see the Table of Analysis of Examination Syllabuses (p. ix) to find out the exact percentage for your examining group).

## Why have internal assessment of practical work?

There are certain skills in Physics which it is important to acquire which are better tested in this way than by conventional written examination papers. Such skills include the ability to:

1 set up and use apparatus safely to make just the measurements asked for,
2 make accurate observations in the laboratory,
3 record results and observations accurately, systematically and economically in appropriate forms (such as tables of results, diagrams and written reports),
4 present results as graphs,
5 design and carry out experiments to solve a problem.

## How does the internal assessment work?

Although details of how the assessment is carried out vary slightly from one examining group to another, the basic arrangements are the same for all groups and all syllabuses.

The assessment is carried out by your physics teacher at certain times during the two years before you sit the written examination and most of the assessment is likely to happen in the last year of your course rather than earlier.

Your teacher has to assess you on a number of practical skills (generally between four and six) and has to test each at least twice. He or she will generally do this during the normal practical work which is part of your physics course. He or she will assess each candidate individually.

Each time your teacher assesses you for a particular skill he or she awards a mark and records the skill and the mark on your personal record sheet. Towards the end of your course the teacher submits these marks to the examining group and also any of your work they may require as evidence for these marks. On the basis of this at least 20% of your marks for the whole examination are calculated.

## What safeguards are there?

The following safeguards ensure that you obtain a fair mark for your practical ability:

1 Each skill is assessed more than once.
2 If your school has a large entry for the examination (more than one set or form) the school has to ensure that the same standards are applied to all candidates. This normally means that those who teach the forms or sets meet to discuss their assessments and adjust their marks if necessary to achieve a common standard.
3 The marks and samples of work of some candidates are sent to someone outside the school (a moderator) who ensures that your school is applying the same standards in their assessment as other schools are. A few examining groups send moderators to schools to help this standardization.

## How can I best improve the standard of my practical work?

The best advice is always to:

1 carry out practical work thoughtfully and sensibly,
2 consider safety.
3 ask yourself if you have set up the apparatus in the most sensible way and the way in which it best enables you to make the measurements or observations required,
4 read measuring instruments carefully and accurately, checking a second time if you are in doubt,
5 check readings that seem wrong,
6 write down your results in your book immediately and clearly, drawing up a table of results if appropriate,
7 draw any necessary graphs neatly in pencil using simple scales on the axes (which should be labelled with both the quantity measured and its unit, e.g. time(s)), making the graph as large as the sheet of graph paper will sensibly allow,
8 write a clear account of what you have done, being careful to explain exactly how you made any measurements,
9 write a short conclusion at the end saying what you think your observations or measurements mean.

# INDEX

Acceleration 3–8
  uniform 4–8
Action 12–13
α-rays 105, 107
Alternating current 92
Ammeter 81
Ampere, the 80, 85
Amplifier-current 100–1
Amplitude of a wave 50, 72
Apparent depth 61–2, 67
Area 1
Atomic structure 104, 107
Atoms 29, 104, 107

Balance, spring 9
Barometer
  aneroid 25–6
  simple 25, 26
β-rays 105, 107
Boyle's law 36–7
Brownian motion 30–1, 32

Camera 57–8, 66
Cathode rays 96–7
Cathode ray oscilloscope 97–8, 102
Centisecond timer 16
Centre of mass 17–18
  location of 18
Charles' law 36, 37–8
Circular motion 11–12, 15
Cloud chamber 103–4
Conduction of heat 47, 49
Conductivity 47
Conductors, electrical 77, 79, 98, 102
Convection 47, 49
Coulomb, the 80
Critical angle 62–3
Current, electrical 80, 85, 98
  induced 91
Current amplifier 100–1

Deflection tube 97, 102
Density 21–2
  measurements 21–2
Diffraction
  of light 68, 70
  of waves 54, 55
Diffraction grating 69, 70
Diffusion 31, 32
Digital electronics 101, 102
Diode 95–6, 102
  light emitting 100, 102
  p-n junction 99, 102
  semiconducting 99
Distance travelled 5
Dynamo
  a.c. 92
  d.c. 91

Earthing 78, 85
Echoes 72, 73
Eclipses 57, 59
Elasticity 27–8
Electric
  bell 87
  current 80
  motor 89–90
Electromagnet 87, 90

Electromagnetic
  induction 91–4
  spectrum 69
Electromagnetism 86–90
Electromotive force 80
Electron
  beams 95–102
  gun 96, 102
Electronic systems 101, 102
Electroscope 77–8
Electrostatics 77–9
Energy 14
  electrical 82–3, 85
  kinetic 14, 16
  nuclear 106–7
  potential 14, 15
Evaporation 29, 46
Expansion 33–5
  of liquids 34–5
  of solids 33, 35
Eye, the 66–7

Faraday's law 91
Fine beam tube 88–9, 97
Fission 106–7
Focal length 63, 64, 67
Focal point 63, 64, 67
Force(s) 9–12, 15
  between molecules 27–8
  centripetal 11–12, 15
  magnetic lines of 76
  on moving charges 88–9, 90
  turning 17
Frequency
  of light 50
  of sound 72
Fuses 84–5

Galvanometer 90
γ-rays 69–70, 103, 105, 107
Gases 36–41
  laws 36, 37, 38, 39
Gates, logic 101–2
Geiger tube 103

Induction
  charging by 78, 79
  electromagnetic 91–4
Inertia 10
Infrared radiation 48, 69–70
Insulators, electrical 77, 79, 98, 102
Intensity of sound 72, 73
Interference
  of light 68, 70
  of waves 53, 55
Isotopes 104–5

Joule, the 13, 16, 83

Kilowatt 15
Kilowatt-hour, the 83
Kinetic energy 14, 16

Latent heat 45–6
  of fusion 45, 46
  of vaporization 45, 46

Length 1
Lenses 63–7
Lenz's law 91
Levers 19, 20
Light-dependent resistor 99, 102
Light emitting diode 100
Logic gates 101–2
Loudness 72

Machines 18–20
Magnetic field 75–6, 86–7
Magnetization 75
Magnets 74–6
Manometers 24–5
Mass 1
  centre of 17–18
Materials,
  semiconducting 98–9, 102
Matter 29–30, 32
Mean free path 31–2
Microscope, the 66
Millisecond timer 1, 6
Mirrors 58–9
Molecules 29
Moments 17, 20
Momentum 12–13, 15
Motion 2
  Brownian 30–1, 32
  laws of 9, 15
  in a circle 11–12, 15
Moving coil meter 90

Newton's Laws of Motion 9, 15
Nuclear energy, fission 106–7

Ohm, the 80, 85
Ohm's law 80, 85

Pitch 72, 73
Poles of magnets 74, 76
Potential
  difference 80, 85
  energy 14, 15
Power 15, 16
  electrical 83, 85
  of a lens 63
  transmission 93–4
Pressure 23–6
  atmospheric 24–6
  in a liquid or gas 23–4, 26
  law 36, 38–9
Principal focus 63
Projectiles 7, 8
Propagation of light 56–8
Pulleys 19, 20

Radiation 47–8, 49
  detectors 103–4, 107
  infrared 48, 69–70
  ultraviolet 69–70
Radioactive decay 105–6, 107
Radioactivity 105
Randomness 31–2
Ray
  box 56
  diagrams 64–5

Reaction 12–13
Rectifier 96, 99
Reflection 52, 55, 58–9
   internal 62–3
Refraction 52–3, 55, 60–7
   at plane boundary 60–2
   at spherical boundary 63–4
   laws of 60
Refractive index 61, 67
Resistance 80–1, 85
Resistors
   in parallel 82, 85
   in series 81–2, 85
Ripple tank 50–2
Root mean square voltage 92–3

Safety
   electrical 84–5
   radioactive 106
Scalar 15, 16
Scaler 103
Second, the 1
Semiconducting materials 98–9, 102
Shadows 56–7, 59

Sound 50, 54, 71–3
Specific heat capacity 42–4
   measurements of 42–4
Spectrum
   electromagnetic 69
   visible 62
Speed 3, 7
Spring balance 9
Stability 18
Stroboscope 1, 7
   hand 1, 7, 51–2
   motor 1, 7

Thermionic emission 95
Thermistor 81, 99
Thermometers 34–5
Ticker-timer 1, 2, 7
Time 1
Timer
   centisecond 1, 6
   millisecond 1, 6
Transformer 93, 94
Transistor 100–1, 102
Truth table 101–2

Ultraviolet radiation 69–70
Universal gas law 39–40

Vector 15, 16
Velocity 3, 7
   of light 50
   of sound 72
Volt, the 80, 85
Voltmeter 81
Volume 1

Watt, the 15, 16, 83
Waves 50–5
   diffraction of 54, 55
   interference of 53, 55
   reflection of 52, 55
   refraction of 52–3, 55
Wavelength 50, 72
Weight 10
Work 13, 15, 18–19, 20

X-rays 69–70, 103